The Complete Guide to Physical Security

The Complete Guide to Physical Security

Dr. Paul R. Baker, CPP • Dr. Daniel J. Benny, CPP

CRC Press is an imprint of the
Taylor & Francis Group, an **informa** business

CRC Press
Taylor & Francis Group
6000 Broken Sound Parkway NW, Suite 300
Boca Raton, FL 33487-2742

© 2013 by Taylor & Francis Group, LLC
CRC Press is an imprint of Taylor & Francis Group, an Informa business

No claim to original U.S. Government works

Printed in the United States of America on acid-free paper
Version Date: 2012928

International Standard Book Number: 978-1-4200-9963-8 (Hardback)

This book contains information obtained from authentic and highly regarded sources. Reasonable efforts have been made to publish reliable data and information, but the author and publisher cannot assume responsibility for the validity of all materials or the consequences of their use. The authors and publishers have attempted to trace the copyright holders of all material reproduced in this publication and apologize to copyright holders if permission to publish in this form has not been obtained. If any copyright material has not been acknowledged please write and let us know so we may rectify in any future reprint.

Except as permitted under U.S. Copyright Law, no part of this book may be reprinted, reproduced, transmitted, or utilized in any form by any electronic, mechanical, or other means, now known or hereafter invented, including photocopying, microfilming, and recording, or in any information storage or retrieval system, without written permission from the publishers.

For permission to photocopy or use material electronically from this work, please access www.copyright.com (http://www.copyright.com/) or contact the Copyright Clearance Center, Inc. (CCC), 222 Rosewood Drive, Danvers, MA 01923, 978-750-8400. CCC is a not-for-profit organization that provides licenses and registration for a variety of users. For organizations that have been granted a photocopy license by the CCC, a separate system of payment has been arranged.

Trademark Notice: Product or corporate names may be trademarks or registered trademarks, and are used only for identification and explanation without intent to infringe.

Library of Congress Cataloging-in-Publication Data

Baker, Paul R., 1958-
The complete guide to physical security / Paul R. Baker and Daniel J. Benny.
pages cm
Includes bibliographical references and index.
ISBN 978-1-4200-9963-8 (hardback)
1. Business enterprises--Security measures. 2. Security systems. 3. Computer security. 4. Private security services. I. Benny, Daniel J. II. Title.

HV8290.B34 2013
658.4'7--dc23
2012023619

Visit the Taylor & Francis Web site at
http://www.taylorandfrancis.com

and the CRC Press Web site at
http://www.crcpress.com

Contents

Preface xv
The Authors xvii
The Contributors xix

1 Physical Security Planning 1
   Introduction 1
   Setting the Stage 3
   Site Planning 4
   Reviewing the Physical Design 7
   References 9

2 Vulnerability Assessments 11
   The Security Survey 11
   Vulnerability Assessment 13
   Developing a Vulnerability Assessment 16
   Understanding the Threats 19
      Natural Threats 21
      Man-Made Threats 23
   Making Adjustments 23
   References 24

3 Security Design and Planning 25
   Security Planning and Evaluation 25
   Security Design Concepts 30
   Security Technologies 35
   EASI Model 36
   Adversary Sequence Diagram 36
   Working with Architects 37
   Working with Contractors 38
      Have Basic Knowledge of Your World 38

Know What You Want 38
Develop a Relationship 39
Budget 39
Construction Review 40
References 41

4 Security Construction Projects 43
New Construction 43
Initial Point of Access 46
Roadway Design 46
Parking 48
Parking Garages 50
Open Area Parking 52
Loading Docks 53
Signage 54
Retrofitting 55
References 57

5 Protection in Depth 59
Protection-in-Depth Concepts 59
Protection Plans 62
Evacuation Drills 64
Incident Response 68
Penetration Tests 70
Access Control Violation Monitoring 75
References 77

6 Perimeter Protection 79
Crime Prevention through Environmental Design (CPTED) ... 79
Protecting with CPTED Concepts 82
Barriers 86
Fences 87
Gates 88
Walls 89
Lighting 90
Types of Lighting Systems 91
Types of Lights 92
Infrared Illuminators 94
References 95

7 Access Control 97
Access Control..97
Anti-Passback 99
    Card Types..100
Access Control Head End 100
    Receptionist 101
    Escort and Visitor Control 101
    Guards 103
Turnstiles and Mantraps...105
References 108

8 Physical Protection Systems 109
Doors..109
    Door Locks 110
        Electric Locks..110
        Electric Strikes..110
        Magnetic Locks 110
Windows 111
    Types of Glass 112
    Glass-Break Sensors 112
Interior Intrusion Detection Systems..113
    Balanced Magnetic Switch (BMS) 114
    Motion-Activated Cameras 114
    Acoustic Sensors..115
    Infrared Linear Beam Sensors 116
    Passive Infrared (PIR) Sensors 116
    Dual-Technology Sensors...117
Perimeter Intrusion Detection Systems 118
    Infrared Sensors 119
    Microwave Systems 119
    Coaxial Strain-Sensitive Cable...120
    Time Domain Reflectometry (TDR) Systems............................120
    Video Content Analysis and Motion Path Analysis 120
References 121

9 CCTV and IP Video 123
STEVE SURFARO
The Essential Guide to Video Surveillance 123
Video Surveillance Use Cases..124

Video Surveillance Impacts Many Other Industries as Its Use Expands 125
Video Surveillance System Classifications 126
    Analog Video Systems Overview 127
    IP Video Systems Overview 128
Image Capture Video Sources—Cameras 130
    Analog Video Cameras 130
    IP Video Cameras 131
    Network Video PTZ Cameras 131
    IP Video Encoders 132
        Compression Technology Overview 133
    License Plate Capture (LPC) Cameras 134
    Cameras with "True" Day/Night Capability 134
    Cameras with Non-IR Sensitive Day/Night Capability or "Chroma Mode Capability" 135
    Thermal Cameras 135
    Progressive Scan CCD 137
Lenses 137
    Varifocal Lens 137
Lighting 137
    Color Temperature 138
    Infrared Illumination 138
    Pixels, Imager Sizes, and Sensitivity 138
HDTV 138
    HDCCTV 141
    HDTV Deployment Justification 143
Recording Systems 143
    Digital Video Recorders 143
    Network Video Recorders 144
Video Management Systems 144
    Video Surveillance and Cloud Computing 144
        The Solution 145
        Public Cloud 147
        Private Cloud 147
        Community Cloud 147
        Hybrid Cloud 148
Video Control, Analysis, and Video Content Analysis Systems 148
    Video Analytics and Automated Object Detection 148
    Content Analysis 149

  Intelligent Video Search 150
  Video Synopsis 150
  Interoperability 151
  Integration 152
  Application Programming Interface 152
 Using a Step-by-Step Approach to System Selection and Deployment. 152
 Upgrade Path 159
 Manpower and Video Surveillance 160
 Privacy and Legal Considerations—Video Surveillance Checklist 162
 The Future of Video Surveillance 163
 References 164

10 Keys, Locks, and Safes 165
 Keys, Locks, and Safes 165
  Locks 166
  High-Tech Keys 168
  Key Control 169
 Developing a Master Locking System 169
 Safes 171
 Vaults 173
 Containers 174
 Reference 174

11 Biometrics 175
ROQUE SOLIS AND CHUCK WILSON
 Introduction 175
 What Is Biometrics? 176
  Facial Recognition 177
  Fingerprint 178
  Hand Geometry 180
  Iris Recognition 181
  Vein Pattern Recognition 182
 Multibiometric Systems 184
 Deploying Biometrics 186
 Privacy 190
 Biometric Metrics 190
  Failure to Enroll 191
  Failure to Acquire 191
  FRR/FAR 192
 Attacks on Biometric Systems and Their Remedies 194

Summary 197
Notes 197

12 Security Guard Force 201
Establishing a Security Guard Force 201
Mission and Duties of the Security Guard Force 202
Profile of the Facility to Be Protected 204
    Mission of Facility 204
    Security Threat 205
    Size of Facility 205
    Hours of Operation 205
    Number of Employees/Visitors 205
Proprietary Security Guard Force 206
Contract Security Guard Force 208
Hybrid Force 209
Security Guard Uniforms 210
Security Guard Identification 211
Security Guard as an Authority Figure 212
Security Guard Protective Equipment 212
    Armed or Unarmed? 213
Security Guard Training 214
Professional Security Certifications 215
Personnel Issues 216
Summary 218
References 218

13 Central Station Design 219
Developing an Operation 219
Design Requirements 221
Secondary Amenities for an Operations Control Center 223
Alarm Assessment 223
References 224

14 Government Security 227
SCIF 227
Location 228
Design 228
    Perimeter Walls 228
    Doors 228
    Windows 228

Electrical and HVAC 230
Sound Masking 231
    Sound Classification 232
TEMPEST 233
Shielding 234
Filtering 234
Access 234
Security Alarm Requirements 235
Open Storage 236
Closed Storage 237
Dealing with Contractors 237
References 238
Appendix A: Fixed Facility Checklist 239

15 Financial Institutions and Banks 247
Introduction 247
Vaults 247
Safes 248
ATMs 249
Night Depositories 250
Teller Cash Recyclers 251
Dye Packs 251
Barriers 252
Ambush Features 253
Video Systems 253
Holdup Alarms 254
    Holdup Buttons 254
    Holdup Foot Rails 255
    Money Clips 255
Bank Guards 255
References 257

16 Data Center Protection 259
Data Centers 259
Fire Protection 263
Fire Detection/Alerting 263
Fire Suppression 265
References 267

17 Total System Cost 269
Determining Total System Cost 269
System Design Cost 270
System Installation Cost 271
System Operational Cost 271
IT-Related Cost 272
Maintenance Cost 273
Replacement Cost 273
Cost-Benefit Analysis 273
Cost of Loss 274
Cost of Prevention 274
Return on Investment 274
Cost Factors 275
References 278

18 Security Master Plan 279
TIMOTHY GILES
Security Master Plan Strategy 279
Engaging the Stakeholders 280
What Should Your Security Philosophies Be? 282
    Contract Security Relationship 283
    What Should Your Security Strategies Be? 283
Technology Migration Strategy 288
Equipment Replacement Schedules 290

19 Security Foresight 293
Introduction 293
Strategic Foresight 294
Foresight Techniques 295
    Mind Mapping 295
    Environmental Scanning 296
    Future Mapping 298
References 304

20 Security Leadership 305
Introduction 305
What Is Leadership? 305
Purpose of Leadership 306
Effective Leadership 307
Why Security Leaders Are Important 308

Understanding the Basics 308
Are Leaders Born or Made? 310
Good Leadership 311
Bad Leadership 312
Going from Bad to Good 316
Forging the Future 318
New Definitions for Leaders in the Twenty-First Century 320
References 322

Index **325**

Preface

As the physical security field continues to grow, there is a requirement to stay up to date on the constant technical enhancements. There is also a need to train the newly recruited members of the physical security trade and make them aware of the nuances of this field. Physical security is the use of multiple devices, hardware, and technology; this pulls into place access control, closed-circuit tv (CCTV), intrusion detection, and environmental controls. The knowledge base for our field is partly technical and partly physical. We have received this training through years of working in the field and collaborating with other members of the physical security community.

The physical security professional needs a clear understanding of the dynamics of his or her organization and the market forces that drive the organization. Being a physical security professional does not permit you to be blind to the economic and strategic forces that engage your operation, business, or company on a daily basis. Leading a security organization is much more than "gates, guns, and guards"; it takes knowledge to generate the confidence and dedication that is necessary in a turbulent world.

The development as a leader within the security field requires the implementation of a sound security strategic plan. This is sometimes called a security master plan. This plan will require a physical security professional to expand his or her thinking beyond security requirements and will require an extensive understanding of the dynamics in the marketplace, employee issues, and management goals. This will then be coupled into a developmental agreement between security and the business side of the house in order to develop a structured security plan.

The Complete Guide to Physical Security will address each topic from the standpoint of a commonsense understanding while at the same time bringing to light new and interesting concepts within the physical security field. There is a need to address the ever changing world of physical security and to understand the marriage between technology and physical hardware. This book will achieve both while serving as a reference guide to all things within the physical security realm.

The Authors

Dr. Paul R. Baker began his security career in the US Marine Corps as a military policeman. He served with the Presidential Helicopter Squadron (HMX-1), Hawaiian Armed Services Police as liaison to the Honolulu Police Department, and as a base MP at Camp Pendleton, California. Upon discharge, he joined the Maryland State Police and worked in all phases of law enforcement, spending the majority of his career in the intelligence and narcotics areas. Upon retirement in 2001, he began the security management phase of his journey, working for the MITRE Corporation, Institute for Defense Analysis, Capital One Bank, and, finally, with the RAND Corporation.

Dr. Baker is board certified in security management as a certified protection professional (CPP) by ASIS (American Society for Industrial Security) International. He holds a doctorate of strategic leadership from Regent University, master of science in criminal justice from Troy University, master of business administration from Baker College, and a bachelor of science in administration of justice from Thomas Edison State College.

Dr. Baker is an adjunct professor for University of Maryland University College in the homeland security field and an adjunct professor for Southwestern College in its security management curriculum.

He is a contributor to the CRC Press books *Official (ISC)²® Guide to the CISSP® CBK®*, 2nd edition and *Official (ISC)² Guide to the ISSAP CBK*.

Dr. Daniel J. Benny is the security discipline chair at Embry-Riddle Aeronautical University Worldwide. He holds a doctor of philosophy in criminal justice from Capella University, master of aeronautical science from Embry-Riddle Aeronautical University, master of arts in security administration from Vermont College of Norwich University, bachelor of arts in security administration from Alvernia College, associate in arts in both commercial security and police administration from Harrisburg Area Community College, and diploma in naval command and staff from the US Naval War College.

He is board certified in security management as a certified protection professional (CPP) and board certified as a professional certified investigator (PCI), both by ASIS International; certified fraud examiner (CFE) by the Association of Certified Fraud Examiners; certified confidentiality officer (CCO) by the Business Espionage Controls and Countermeasures Association; certified member (CM) airport certified employee-security (ACE); and airport security coordinator by the American Association of Airport Executives. He is a licensed private investigator and security consultant in the Commonwealth of Pennsylvania and holds a black sash in Jeet Kune Do.

Dr. Benny is the author of the CRC Press books *General Aviation Security: Aircraft, Hangars, Fixed-Base Operations, Flight Schools, and Airports*; *Cultural Property Security: Protecting Museums, Historic Sites, Archives, and Libraries*; and the forthcoming book, *Industrial Espionage: Developing a Counterespionage Program Security*. He has authored over 300 articles on security administration, intelligence, aviation security, private investigation, and cultural property security topics.

He served as a US naval intelligence officer with duty at the Office of Naval Intelligence, Naval Investigative Service, Willow Grove Naval Air Station, Fleet Rapid Support Team, and Central Intelligence Agency. He also served as Director of Protective Service Pennsylvania Historic and Museum Commission, and as US Navy police chief.

The Contributors

Timothy Giles is the president of Risk/Security Management & Consulting. Prior to going into business for himself, Mr. Giles was the managing director of security services for Kroll Associates in Atlanta for 2.5 years. Before that, he served as director of security for North America at IBM where he was the executive responsible for the firm's security operations for all of the United States and Canada until he retired after 31 years of service. Mr. Giles has also worked in IBM's Latin America operations as the director of security for 2 years and spent 3 years living and working in the Asia Pacific Region of IBM's security operations as the area security manager. While in Asia, he was responsible for the security planning for IBM during the 1988 Olympics in Seoul, South Korea. In previous careers at IBM, Mr. Giles worked in manufacturing and engineering positions in IBM's semiconductor operations.

Mr. Giles became board certified in security management as a certified protection professional (CPP) by ASIS International in 1997 and as a physical security professional (PSP) in 2007. He was selected as Security Director of the Year in 1997 by *Access Control & Security Systems Integration* magazine. During his more than 25 years in security, he has worked in all aspects of physical security, information protection, investigation, crisis management, emergency planning and response, and disaster contingency planning, as well as managing major projects in most of these areas. Mr. Giles has become accomplished in utilizing all aspects of security technology. He spent 18 years working in the corporate security arena and has spent the following years working as a security consultant. He is also a licensed private investigator.

Roque Solis is an experienced IT professional and consultant with over 20 years of successful engagements, both public and private, from new product development to improvements of existing solutions. Roque Solis is an internationally recognized security expert in biometrics, PKI (public-key infrastructure), and integrated circuit chip technology. Mr. Solis has designed cutting edge biometric solutions of various modalities for both physical and logical security applications. He created a smart chip-based solution that supported a set of shared cryptographic keys in POS terminals for multiple state governments. He implemented a highly successful biometric solution for a large automotive manufacturer. Most recently, Mr. Solis has engineered an exceptional password management solution that is in use throughout the world. He can be reached via his company's Web site: http://www.solisystems.com/

Steve Surfaro is an Axis Communications Business Development manager and security industry liaison and has more than 25 years of security industry experience. He works with systems integrators, data transport professionals, consultants, and end-users to achieve successful IP and SaaS (online portal) video applications. Mr. Surfaro has spoken at numerous industry events, including ISC, ASIS, and BICSI. He is a member of the ASIS Physical Security Council and speaks regularly at their workshops. He is the chairman of that council's education subcommittee and is responsible for consistent workshop content. He is regular contributor to BICSI, as well as SIA's Digital Video Standards.

Mr. Surfaro's specialties include integrated security system design, command center design, deployment of IP Video Solutions, speaker and educator at industry public events and Webcast seminars. Specific experience includes video surveillance for security, gaming surveillance, physical and logical access control, fire alarm/life safety, and mass communications systems. He has additional certifications with Addressable Fire Alarm, Voice Evacuation, Floor and Floor Above Signaling, PACS Fail-Safe Shutdown, and Fan Shutdown & Override Systems. Mr. Surfaro has been quoted and published in many infrastructure, safety, and security-related publications such as *Access Control, Security Director News*, and *Security Products*. He holds a BA degree in engineering from The Cooper Union and resides in New York.

Chuck Wilson is a published, global authority on smart cards and biometric solutions. Mr. Wilson has been in the information technology industry for over 30 years, and has been researching and writing about enabling technologies such as smart cards and biometrics for the past 16 years. Mr. Wilson manages Smart Solutions at Southlake, TX-based ii2P, a company that has pioneered self-service IT solutions.

He has written or co-written three previous books: *Get Smart*, 2001, regarding the emergence of smart cards in the USA; *RFID: Item Level Management*, 2005, and *Vein Pattern Recognition, A Privacy-Enhancing Approach*, published by CRC Press in 2010. He received his BA and MA degrees from The Ohio State University, and his MBA from Memphis State University.

Chapter 1

Physical Security Planning

Paul R. Baker

Introduction

From the physical security standpoint, we are the gatekeepers. We are the protectors of the organization from all threats; regardless of whether they are malicious, internal, or environmental. We need to be ever vigilant and confident that the work we are doing will be regarded as a necessary operational function for the overall security and the protection of the asset—to include employees, information, and property. Protection of these three things is the cornerstone of our profession (Figure 1.1).

Much of physical security is just common sense, but there are some twists and turns coupled with a few nuances that need to be clarified. There are a multitude of concepts, ideas, and theories from different vocations ranging from criminology to architecture on how to reduce the incidence of crime and the protection of a facility along with significant advances in physical security technology. Within each area of security there are ways to explain these concepts and understand the basic principle of each theory. For the average security professional, the best way is the simplest and most direct way in order to get the point across and make it understandable. This is not a technical approach to physical security but rather a hands-on and learning environment that will provide the reader with an overall understanding of what works and what does not.

According to the *Dictionary of Military and Associated Terms,* physical security is defined as the part of security concerned with physical measures

Figure 1.1 The Queen's Guards at Buckingham Palace in England have guarded regents and the royal palace since the year 1660.

designed to safeguard personnel; to prevent unauthorized access to equipment, installations, material, and documents; and to safeguard them against espionage, sabotage, damage, and theft.[1]

Physical security is often described as the "forgotten side of security" and yet it is a key element to an overall protection strategy. Protection of restricted work areas is important to the overall functionality of the company operation. Proprietary, sensitive, and classified material must be protected from the general population or from other employees who do not have the need to know. In the case of a company, such areas may be protected by restricting unauthorized personnel from entering the area. General traffic flow to the area must be diverted away to minimize the entry of unauthorized personnel. Personnel who are authorized to enter these restricted areas must have a company badge that quickly identifies them as authorized. Moreover, these authorized personnel must be on an access roster that the guards or access control system can use to verify their credentials. If the area is a large area where vehicles are used, guards at the sentry point would have to verify the vehicles being used to enter the premises along with personal identification. Verification may be accomplished by using access rosters for both vehicles and personnel. By applying these preventive measures, the risk of loss or damage is reduced.

Many organizations spend thousands of dollars on IT hardware and software—only to forget about securing the actual building that houses them. Remember: Even if no one can steal or corrupt your data over the network, someone may still be able to walk out your front door with that information. Do not neglect physical security in your attempts to lock down data.

Setting the Stage

In incorporating a physical security survey it is necessary to begin the project by looking at it from a 10,000 foot level. In all endeavors within security, the concept of having multiple eyes on a project is beneficial. Selecting a team is one of the first requirements for any security project and it will be essential to have different perspectives to ensure that all areas of security have been covered.

The security professional has extensive responsibilities toward the organization in terms of protecting information, personnel, and property. Within this venue it is up to the security professional to challenge the norm and produce ideas that will effectively protect the company and develop the security operation. "To accept a challenge means to accept responsibility for generating ideas as possible solutions to the problem. The more you accept responsibility and dedicate yourself to generating ideas, the higher your probability of reaching an innovative solution."[2] This is an example of how creativity is essential in developing a comprehensive security initiative.

A physical security program is designed to prevent the interruption of operations and provide for the security of information, assets, and personnel. Operational interruptions can occur from natural or environmental catastrophes like hurricanes, tornadoes, and floods as well as from industrial accidents like fires, explosions, or toxic spills; intentional acts of sabotage; vandalism; or theft.

During the design phase of a site, it should be the standard operational procedure for a security professional to review all aspects of construction to include: land use, site planning, stand off distance, controlled access zones, entry control, vehicular access, signage, parking, loading docks, service access, security lighting, and site utilities. Integrating security requirements into a comprehensive approach necessitates achieving a balance among many objectives, such as reducing risk, architectural aesthetics, creating a safe work environment, and hardening of physical structures for added security.

It is important to remember that the nature of any threat is always changing. The great Yankee catcher Yogi Berra once said, "If you don't know where you are going, you might wind up someplace else." In other words, having some direction is your best approach in dealing with physical security. Do not be reactive and start throwing money into security systems. This was evident after 9/11 when security budgets ballooned. Be insightful in identifying necessary security needs and put a plan together along with a budget in order to achieve the overall objective.

The world of security is changing faster than we sometimes realize or wish to acknowledge. In addition to the age-old problems of violence and crime, security professionals now contend with international terrorism, environmental damage, energy disruptions, and potential pandemics. To these protracted and almost universal problems one can add the prospect of unexpected and often violent natural events like earthquakes, floods, hurricanes, fires, or tornadoes.

Site Planning

The single most important goal in planning a site is the protection of life, property, and operations. Effective building security requires careful planning and design of the physical protection system, integrating people, procedures, and equipment—the foundation of all facility security operations. A security professional will need to make decisions in support of this purpose and these decisions should be based on a comprehensive security assessment of the threats and hazards so that planning and design countermeasures are appropriate and effective in the reduction of vulnerability and risk. Protecting a building, its occupants, and related assets can pose a complex problem, and there is no perfect defense to all of the potential threats a target may face. Optimizing building security with respect to performance, cost, and efficiency ultimately requires compromise and balance in the application and consideration of people, procedures, and equipment.

There is a natural conflict between making a facility as convenient as possible for operation and maintaining a secure facility. If it were only

Figure 1.2 As secure structures go, castles are pretty unbeatable. However, not every building or facility warrants being a fortress.

up to security in designing a facility, it would look like a fortified castle (Figure 1.2).

However, with most applications and design requirements there needs to be cooperation between several departments. Convenience should be considered during the different phases of the design review; however, the requirement for security should never be sacrificed for convenience. In most instances, proper security controls will reduce the flow rate and ease of entry and egress in and out of a facility. These issues must be addressed in the initial planning to facilitate additional entry points or administrative requirements. Once a process has been established and there is buy-in from the employees and management, the acceptance of normal operations is now approved. When there are changes to the design after the fact and personnel are used to doing something a certain way, there will be reluctance, questions, and push-back. Humans like stability and have a need for things to be a certain way.

For example, for years an employee entrance was in the rear of the building; security recommended closing it down because of guard cost and lack of use. This entrance was converted to an "emergency exit only" and when this was done and notification was given to the general employee population, even employees who did not use it when it was an entry point wanted to know why this had been done. This is because of *change*; it takes away from the employee norm and becomes a change in their routine. And as we all know, in every company, there is that group of employees that will always be asking why. However, if a sound security explanation is presented, it will not take long for the grumbling to end and the new routine will take its place.

To maximize safety and security, a design team should implement a holistic approach to site design that integrates security and function to achieve a balance among the various design elements and objectives. Even if resources are limited, significant value can be added to a project by integrating security considerations into the more traditional design tasks in such a way that they complement the design.

A company's employees are perhaps the most valuable asset of the business, and other building occupants, such as customers, contractors, vendors, and visitors, must be considered in the overall security design. From a legal and ethical standpoint, the building's owner has a duty to provide protection for all occupants of the building depending on the terms of the lease and the contract with the security company. These people fall into the category of those occupants in a building that need to be protected. At the outset of the security design process, their functions and movement patterns need to be analyzed and incorporated into the design to help determine design requirements for procedures and equipment.[3]

Another category of people that needs consideration are those who provide protection. In the "people" element of security operations, the integration of people as a layer of security in the form of security guards and other security personnel is another consideration. The architectural layout of the building can be designed to influence the movement of people for rapid evacuation, limited congregation, and increased visibility, and to limit the need for an abundant number of security personnel.

Security is a dynamic process that, in order to be effective, must be procedural in nature. The premise is to develop procedures involving such events as evacuation, emergency response, and disaster recovery in response to fires, natural disasters, and criminal intrusions. Emergency response and business continuity plans must be developed well in advance of a critical incident and must define the plan of action, or steps to be taken in a logical, orderly, and procedural manner. Policies and procedures are then developed to assist with the proper response and recovery actions in the event a crisis strikes. It should be noted that policies and procedures are guides to actions that should be taken in the event of an emergency and should not be strictly construed. No two critical incidents are the same, and it would be impossible to develop a procedure for every possible event that could occur. Procedures should be written so that they can be adaptable to any application.

The security elements of people, procedures, and technology are interdependent because they rely on each other to be effective. The behaviors and needs of people dictate what procedures and equipment may be deployed; procedures depend on people to be effective and equipment requirements depend on the particular procedures to be followed in a critical incident. A cost-effective and comprehensive plan necessitates a balance of these three elements taking into account their particular contribution to the mission; one application may be security personnel intensive, whereas another may be equipment intensive.[4]

Reviewing the Physical Design

There is a natural conflict between making a facility as convenient as possible for operation and maintaining a secure facility. Convenience should be considered during the different phases of the design review; however, the requirement for security should never be sacrificed for convenience. Proper security controls will reduce the flow rate and ease of ingress and egress in and out of a facility. These issues must be addressed in initial planning to facilitate additional entry points or administrative requirements. As a security professional, you need to be passionate about controlling access to the individuals who need to be inside the protected space. This needs to be done with a robust access control system; however, having a system in place also requires the employees to embrace the system and utilize it with the assurance that it will provide secured access with limited interruption.

To maximize safety and security, a design team should implement an all-inclusive approach to site design that integrates security and function to achieve a balance among the various design elements and objectives. Even if resources are limited, significant value can be added to a project by integrating security considerations into the more traditional design tasks in such a way that they complement the design.

The movement of people and materials throughout a facility is determined by the design of its access, delivery, and parking systems. Such systems should be designed to maximize efficiency while minimizing conflicts between entry and exiting of vehicles and pedestrians. Designers should begin with an understanding of the organization's requirements based on an analysis of how the facility will be used.

Figure 1.3 It is essential to visit and examine facilities and facility plans for buildings in development to understand inherent functionality, business operation, and security issues.

The need to review the existing facility or facility plan (Figure 1.3) in order to learn all of the operations, conditions, and features that affect the overall security requirements is paramount. Here, you as the security professional or a member of the security assessment team need to walk and view the areas that require protection. This is when you want to do your on-site interviews with personnel who will be working inside the facility. Do not take it for granted that nonsecurity persons do not know or understand the conditions required for their own protection. Look at it from the standpoint of a thief: What is valuable? What would a thief want? How would a thief obtain it?

Start with a list of the assets and determine what the threats are. Look at each entry point and determine the vulnerabilities and check potential penetration points and blind spots—look at it as if your life depended on getting in and how you would get in. After putting yourself in the shoes of a perpetrator, begin to reverse engineer and think of ways to defeat this person from entering.

The idea is to walk a site physically during normal business hours and after hours. Talk with personnel during normal business hours and after hours. If there is a midnight security guard, this person will be a treasure trove of information. He or she has nothing but time on his or her hands and walks the facility without interruption seeing things that are only clearly visible at night. Walk the site with a team and place more eyes on the project, which can only benefit you when assessing a site. This is a very important point that I want to make crystal clear: Security is a team effort. The more eyes, input, and collaboration there are, the better the final product will be.

When this initial phase is completed, it is necessary to sit with all stakeholders and understand their needs and requirements. Let them explain how they would like the flow of employees, guests, and deliveries to proceed during a normal business day. This will allow you as the security professional to take the information and massage it into a security site plan that will incorporate the stakeholders' needs within a protective envelope. This will give you the opportunity to provide the stakeholders with a plan with security and safety in mind in order to progress toward the next steps in the site planning stage and the development of policy and procedures that all parties can present to the employee base as the required steps in the protection of and accessibility to the facility.

References

1. *Dictionary of Military and Associated Terms.* 2005. US Department of Defense.
2. Michalko, M. 2006. *Thinkertoys: A handbook of creative-thinking techniques,* 2nd ed. Berkeley: Ten Speed Press.
3. Garcia, M. L. 2006. *Vulnerability assessment of physical protection systems.* Burlington, MA: Elsevier Butterworth-Heinemann.
4. Kovacich, G. L., and E. Halibozek. 2003. *The manager's handbook for corporate security: Establishing and managing a successful assets protection program.* Burlington, MA: Elsevier Butterworth-Heinemann.

Chapter 2

Vulnerability Assessments

Paul R. Baker

The Security Survey

Before any project begins, there must be an assessment made in order to put together an operational plan and a practical approach to securing the facility. This security assessment can also be called a security survey, vulnerability assessment, or risk analysis.

No one with any common sense starts a project without a plan. A ship's captain would never leave port without navigational tools, maps, global positioning systems, and a seasoned crew. The same goes for security professionals who will need the tools to initiate a security assessment. It makes no sense to throw up cameras without any rhyme or reason; this would be a waste of resources and money, both of which are in short supply.

A *security assessment* is a comprehensive overview of the facility including physical security controls, policy, procedures, and employee safety. At the beginning, a good assessment requires the security professional to determine specific protection objectives. These objectives include threat definition, target identification, and facility characteristics.

The first question a security professional should be asking is, "What is the threat?" Then start down the list of the potential threats to the organization or facility. Is it vandals, hackers, terrorists, internal employees, corporate spies, or a combination? Stating the threat will identify how adversaries can impact assets and will provide the guidance to developing a sound physical protection system.

Table 2.1 Sample of Defined Threat Matrix

Asset	Probability of Attack	Consequence of Loss
Data center server	Medium	Very high
Portable laptops (critical staff)	High	High
Copy machine	Low	Low
Portable laptops (nonessential personnel)	High	Low
PCU	Medium	High
Classified containers	Low	High

Target identification is concerned with the most valuable asset that needs to be protected. Assets can be personnel, property, equipment, or information. To identify assets to be protected, it would be prudent to prioritize the assets or establish a matrix and identify the asset in conjunction with probability of attack, along with the question: What would be the impact and consequence of the loss of the asset? (See Table 2.1.)

Facility characteristics involve several things to look at from the standpoint of whether this is an existing structure or a new construction. The security professional is reviewing architectural drawings, doing a walk-through of the facility, or both. In the case of an existing structure, it is recommended that a team of security personnel walk the facility. This goes with the old adage that "two heads are better than one." Having several eyes on the project will assist in developing a good evaluation.

The American Institute of Architects has established some key security questions that need to be addressed while performing the security assessment:[1]

1. What do we want to protect?
2. What are we protecting against?
3. What are the current or expected asset vulnerabilities?
4. What are the consequences of loss?
5. What specific level of protection do we wish to achieve?
6. What types of protection measures are appropriate?
7. What are our protection constraints?
8. What are the specific security design requirements?
9. How do the integrated systems of personnel, technologies, and procedures respond to security incidents?

Walking with a team of security professionals through a facility will provide a static presentation of how to protect the facility. However, one of the best ways to build a comprehensive approach toward protecting the facility is by doing on-site interviews. Everyone has an opinion on security and it is amazing that the best insight and information on what needs to be protected and how it should be protected comes from interviewing the staff. As I have stated before, I have found that one of the most astute and insightful persons interviewed has been the midnight security officer. He or she has nothing but time and walks the facility without interruption, seeing things that are only clearly visible at night.

Another effective tool in assessing the current climate of security is to distribute a security questionnaire to gather a baseline for quantifiable data (Table 2.2). Once these questions are answered and a thorough facilities evaluation and staff interview are completed, it is time to develop and outline a physical protection system for the facility.

Vulnerability Assessment

The assessment of any vulnerability of a facility or building should be done within the context of the defined threats and the value of the organization's assets. That is, each element of the facility should be analyzed for vulnerabilities to each threat and a vulnerability rating should be assigned. You would not install $10,000 worth of security equipment in order to protect $100 worth of assets. Also, the decision to implement or not to implement countermeasures may be driven by the importance of the system or its data or by mandates, as opposed to its cost. In either case, the sum of averted risks must be considered where a single remedy will reduce several risks. The security professional must also consider the use and interaction of multiple remedies. One remedy may improve or negate the effectiveness of another.

These considerations form the basis for determining which protective measures are the most appropriate. After having evaluated each risk and the loss that could occur, assessments can be made about the funds that can be allocated to lessen the estimated annual losses to an acceptable level. With information on loss before and after the application of controls, cost evaluations will indicate which countermeasures are most cost effective. When identifying the protective measures that should be implemented, consideration should be given first to the greatest risks.[2]

Table 2.2 A Standard Security Questionnaire

	Strongly Agree	Agree	Neutral	Disagree	Strongly Disagree
The location of the facility is generally safe					
The surroundings of the facility are generally safe					
It is safe to go to the facility anytime, even during the night					
During nighttime, the grounds are protected and secured; the outside lighting makes the facility secure					
It would be difficult for an intruder to gain access to the compound					
The fence around the compound is effective at keeping intruders out					
My vehicle is safe and secure in the parking lot					
It would be difficult for an intruder to gain access to the building through the loading dock					
It would be difficult for an intruder to gain access to the main building					
It would be difficult for an intruder to go undetected inside the main building					
It would be difficult for an intruder to gain access to the corporate offices					
Additional comments:					

Table 2.3 Sample of Vulnerability Matrix

Main Facility	Vulnerability
Front entrance	Medium
Receptionist	High
Access control	Low
Response to alarms	High
CCTV	Medium
Classified containers	Low

It should be noted that a vulnerability assessment (Table 2.3) may change the value rating of assets due to the identification of critical nodes or some other factor that makes the organization's assets more valuable:

Very high. One or more major weaknesses have been identified that make the organization's assets extremely susceptible to an aggressor or hazard.
High. One or more significant weaknesses have been identified that make the organization's assets highly susceptible to an aggressor or hazard.
Medium high. An important weakness has been identified that makes the organization's assets very susceptible to an aggressor or hazard.
Medium. A weakness has been identified that makes the organization's assets fairly susceptible to an aggressor or hazard.
Medium low. A weakness has been identified that makes the organization's assets somewhat susceptible to an aggressor or hazard.
Low. A minor weakness has been identified that slightly increases the susceptibility of the organization's assets to an aggressor or hazard.
Very low. No weaknesses exist.

FEMA (Federal Emergency Management Agency) uses a methodology approach in which the use of an assessment is to achieve the level of protection through mitigation measures with the building design:

- Asset value
  - Identify the criticality of the asset
  - Identify the number of people in the building
- Threat/hazard assessment
  - Identify each threat/hazard

- Define each threat/hazard
- Determine threat level for each threat/hazard
- Vulnerability assessment
  - Identify building and site design issues
  - Evaluate design issues against type of threat
  - Determine level of protection sought for each measure of threat
- Risk assessment
  - Likelihood of occurrence
  - Impact of occurrence
  - Determine relative risk for each threat against each asset

Developing a Vulnerability Assessment

Vulnerability is a condition of weakness. A condition of weakness creates an opportunity for exploitation by one or more threats. The level of risk is determined by analyzing the interrelationship of threats and vulnerabilities. A risk exists when a threat has a corresponding vulnerability, but even high-vulnerability areas are of no consequence if no threats occur. When performing a vulnerability assessment (VA), the primary purpose is to evaluate all the components of the physical protection system. "The key to a good VA is accurately estimating component performance."[3]

When the exploitation of vulnerability occurs, the asset suffers an impact. The losses are categorized in impact areas titled as the following:

1. *Disclosure* is a confidentiality issue. Greater emphasis is placed on this impact area when sensitive or classified information is being processed.
2. *Modification* occurs when an asset is changed from its original state.
3. *Destruction* is when an asset is damaged beyond practical use by threat activity.
4. *Denial of service* is an impact that is emphasized when threats are more likely to cause a temporary loss of capability than total destruction of modification.

Therefore, by emphasizing one or more impact areas in the evaluation process, management can focus their resources on reducing the impact in the area that concerns them most.[4]

The primary goal of a physical protection program is to prevent the interruption of operations and provide for the security of information, assets, and personnel. In this concept, access control into the facility is utilized by barriers arraigned in layers, with the level of security growing progressively higher as one comes closer to the center or the highest protective area. This design requires the attacker to circumvent multiple defensive mechanisms to gain access to the targeted asset:

> Implementing defense in depth requires that the security practitioner understand the goals of security. Essentially, security can be distilled down to three basic elements: availability, integrity, and confidentiality. Availability addresses the fact that legitimate users require resources, which should be available to the users as needed. Integrity relates to the concept that information is whole, complete, and remains unchanged from its true state. Confidentiality can be defined as ensuring that data is available to only those individuals that have legitimate access to it.[5]

Defending an asset with multiple postures can reduce the likelihood of a successful attack; if one layer of defense fails, another layer of defense will hopefully prevent the attack, and so on (Figure 2.1).

For example, consider the layers of security at a local bank, which employs many redundant measures to protect personnel and assets. The

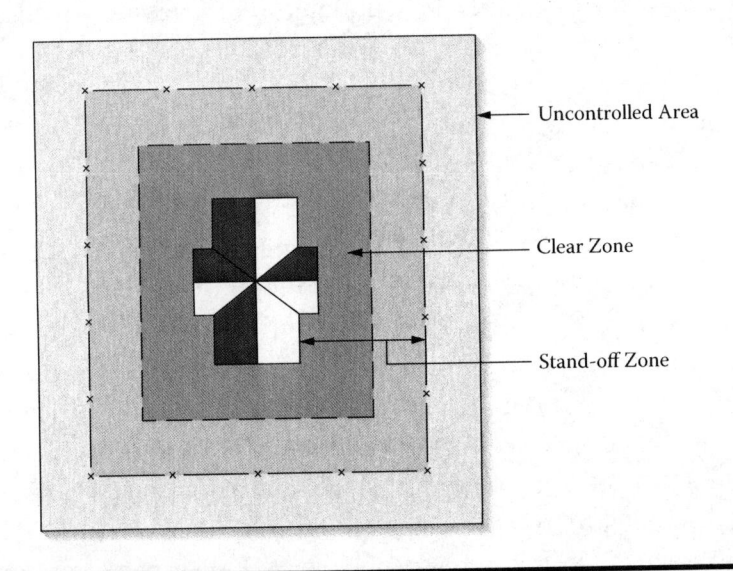

Figure 2.1 Zones provide added layers of defense.

fortress-like appearance and protective reputation synonymous with banking are likely a deterrent factor to some would-be bank robbers, but of course not all. The next line of defense that serves as both a deterrent and as a means for suspect apprehension and asset recovery is security cameras. This layer of security obviously has a level of failure: How many times have we seen video of bank robberies showing a suspect that was never caught? If the cameras are considered ineffective, the next layer is an armed security guard present as a deterrent factor and to defend the bank physically. This, too, is not 100% effective as the security guard can be neutralized by the intruder.

If the security guard is overpowered, the next layer involves hardware, such as bulletproof glass and electronically locked doors. Of course, not all branch offices are fortified in this manner, leaving the bank tellers vulnerable. In this case, the teller must rely on the silent alarm button, dye packs, and robbery training. Some branches also have double time-release doors where people are slightly delayed during ingress and egress.

The vault itself has defense in depth through multiple layers of defense, such as opening only at certain controlled times, its heavy metal construction, and multiple compartments that require further access.

The defense-in-depth principle may seem somewhat contradictory to the "secure the weakest link" principle, since we are essentially saying that defenses taken as a whole can be stronger than the weakest link. However, there is no contradiction; the principle "secure the weakest link" applies when components have security functionality that does not overlap. But when it comes to redundant security measures, it is indeed possible that the sum protection offered is far greater than the protection offered by any single component.[6]

Of course, all of these defenses collectively do not ensure that the bank will never be successfully robbed—even at banks with this much security. If the bad guy wants to rob the bank, he is going to give it his best effort. Nonetheless, it is quite obvious that the sum total of all these defenses results in a far more effective security system than any one defense alone. This does not mean that every known defensive measure should be indiscriminately applied in every situation. Using risk, vulnerability, and threat assessment, a balance has to be found between security provided by the defense-in-depth approach and the financial, human, and organizational resources that management is willing to expend.

The key to a successful system is the integration of people, procedures, and equipment into a system that protects the targets from the threat. A well-designed system provides protection in depth, minimizes the consequences of component failures and exhibiting balanced protection.[7]

Physical protection is no different from computer security and, in fact, it is a dovetail of the processes: You perform a threat analysis, design a system that involves equipment and procedures, and then test it. The system itself typically has a number of elements, which fall into the essence of *deter–detect–delay–respond*.

- **Deter** is meant to render a facility as an unattractive target so that an adversary abandons attempts to infiltrate or attack. Examples of deterrence are the presence of security guards, adequate lighting at night, signage, and the use of barriers, such as fencing or bars on windows. While deterrence can be very helpful in discouraging attacks by adversaries, it cannot stop an adversary who chooses to attack regardless of your defenses—similarly to the bank robber who is absolutely set on robbing the bank and whom nothing is going to stop from attempting to rob the bank. The deterrent value of a true physical protection system can be very high while at the same time providing protection of assets in the event of an attack.
- **Detect** involves the use of appropriate devices, systems, and procedures to signal that an attempted or actual unauthorized access has occurred. It will have one or more layers of barriers and sensors that will be utilized to keep out casual intruders, detect deliberate intruders, and make it difficult for them to defeat your defensive security easily.
- **Delay** involves having a perpetrator delayed by the use of layered defenses. It will delay the attack for a sufficient period of time to allow a response force time to confront and intercept.
- **Response** requires communication to a response force that an unauthorized person is attempting to enter or has entered the facility. The response force is required to intercept the adversary before an attack has occurred or has been completed.

Understanding the Threats

The physical threat to your facility must be defined as part of developing the objectives of the physical protection system. A methodology for defining the threat consists of listing the information needed to define the threat. A list of necessary information might include the type of adversary and possible adversary tactics, potential actions of the adversary, motivations of the adversary, and physical capabilities of the adversary. This is not to say that the threat is always a person or an adversary. A threat can be natural or can be man-made.

There are various sources of information on threat. Intelligence sources can provide detailed information about groups that might pose a threat to your operation. Crime studies that review past and current crimes can provide useful information for characterizing the potential threat. With electronic databases, current published literature can provide extensive information concerning threat. The threat information can then be tabulated and summarized so that adversaries can be ranked in order of their threat potential to a specific facility. Setting up local networks for information exchange, such as meetings of area security professional organizations, can provide information on the assessment of threat. This can be accomplished by reaching out to your counterparts throughout the community.

Having a network of security professionals who can give real-time information on attacks and security issues is extremely valuable. A thief working an area breaking into cars does not want one organization warning another—not only is he watching out for the police but now the organization's security force is also aware of potential break-ins and is on a higher alert for suspicious activity in the parking lots. This information can also be disseminated to the employees to be more aware and make sure they secure their cars (Figure 2.2).

Figure 2.2 Sharing information, whether it be on a rash of car break-ins or other crimes, can help to inform employees, local businesses, and law enforcement—can help be a "force multiplier" and an aid in crime deterrence.

Natural Threats

Natural hazards typically refer to events such as an earthquakes, floods, tornadoes, or hurricanes. Preparation for these natural hazards is accomplished by establishing a communication system in order to get information to employees and upper management. Information and periodic emergency training exercises are the best ways to be prepared and thereby reduce fear and anxiety.

Organizations can also reduce the impact of disasters by flood proofing, installing emergency power systems, and securing items that could shake loose in an earthquake.

From the FEMA "Are You Ready?"[8] series, the following are specific natural threats and ways to deal with each:

> Hurricane is a type of tropical cyclone, the generic term for a low-pressure system that generally forms in the tropics. A hurricane is accompanied by thunderstorms and, in the Northern Hemisphere, a counterclockwise circulation of winds near the earth's surface.
> All Atlantic and Gulf of Mexico coastal areas are subject to hurricanes or tropical storms. Parts of the southwest United States and the Pacific Coast experience heavy rains and floods each year from hurricanes spawned off Mexico. The Atlantic hurricane season lasts from June to November, with the peak season from mid-August to late October.
> Hurricanes can cause catastrophic damage to coastlines and several hundred miles inland. Winds can exceed 155 miles per hour. Hurricanes and tropical storms can also spawn tornadoes and create storm surges along the coast and cause extensive damage from heavy rainfall.
> Hurricanes are classified into five categories based on their wind speed, central pressure, and damage potential. Category Three and higher hurricanes are considered major hurricanes, though Categories One and Two are still extremely dangerous and warrant your full attention.
> Tornadoes are nature's most violent storms. Spawned from powerful thunderstorms, tornadoes can cause fatalities and devastate a neighborhood in seconds. A tornado appears as a rotating, funnel-shaped cloud that extends from a

thunderstorm to the ground with whirling winds that can reach 300 miles per hour. Damage paths can be in excess of 1 mile wide and 50 miles long. Every state is at some risk from this hazard.

Earthquakes are one of the most frightening and destructive phenomena of nature. An earthquake is a sudden movement of the earth, caused by the abrupt release of strain that has accumulated over a long time. For hundreds of millions of years, the forces of plate tectonics have shaped the earth, as the huge plates that form the earth's surface slowly move over, under, and past each other. Sometimes, the movement is gradual. At other times, the plates are locked together, unable to release the accumulating energy. When the accumulated energy grows strong enough, the plates break free. If the earthquake occurs in a populated area, it may cause many deaths and injuries and extensive property damage.

Floods are one of the most common hazards in the United States. Flood effects can be local, impacting a neighborhood or community, or can affect entire river basins and multiple states.

However, all floods are not alike. Some floods develop slowly, sometimes over a period of days. But flash floods can develop quickly, sometimes in just a few minutes and without any visible signs of rain. Flash floods often have a dangerous wall of roaring water that carries rocks, mud, and other debris and can sweep away most things in their path. Overland flooding occurs outside a defined river or stream, such as when a levee is breached, but still can be destructive. Flooding can also occur when a dam breaks, producing effects similar to flash floods.

Be aware of flood hazards no matter where you live, but especially if you live in a low-lying area, near water, or downstream from a dam. Even very small streams, gullies, creeks, culverts, dry streambeds, or low-lying ground that appears harmless in dry weather can flood. Every state is at risk from this hazard.

Man-Made Threats

Threats from fire can be potentially devastating and can affect an organization beyond the physical damage. Not only fire, but also heat, smoke, and water can cause irreversible damage. This type of damage can keep a company from ever regaining its market share and is the leading cause of environmental failures for a company.

The National Archives and Records Administration report states the following:[9]

- Of businesses suffering a fire disaster, 43% never recover sufficiently to resume business. Of those that do reopen, only 29% are still operating 2 years later.
- Of companies that lost their IT (information technology) area for more than 9 days, 93% had filed for bankruptcy within 1 year of the disaster.
- Of businesses that found themselves without their data for more than 9 days, 50% filed for bankruptcy immediately thereafter.

The fire protection system should maintain life safety protection and allow for safe evacuation from the building. A facilities fire protection water system should be protected from a single point of failure. The incoming line should be encased, buried, or located 50 feet away from high-risk areas. The interior mains should be looped and sectionalized. Water can be your main fire suppression tool; however, for electronic equipment, it will cause extreme damage.

Making Adjustments

A good risk management strategy to use when planning a security project consists of assuming that certain aspects of the plan will go wrong and quickly adapting to the new conditions after plan A has failed. For example, underestimating the cost of labor to complete a specific task could result in unexpected expenditures. If the budget is already allocated tightly into all other functions, additional funding will be necessary, unless a special fund was pre-established during the planning phase to cover unexpected expenditures. This is especially important, because once a project and its funding are approved, additional funding will prove difficult to obtain. In most cases a 5% to 10% WAG will be thrown into the approved budget. This is because everyone in the finance and the budgeting group knows there are always overruns and that it is best to prepare and accept this eventuality rather than to go back and ask for additional funding. If you are right on target, then the money is freed up and returned to finance.

References

1. The American Institute of Architects. 2004. *Security planning and design: A guide for architects and building design professionals.* Hoboken, NJ: Wiley Publishing.
2. Chapman, C. 1996. *Project risk management: Processes, techniques and insights.* New York: John Wiley & Sons.
3. Garcia, M. L. 2006. *Vulnerability assessment of physical protection systems.* Burlington, MA: Elsevier Butterworth-Heinemann.
4. Koller, G. 1999. *Risk assessment and decision making in business and industry.* Huntersville, NC: CRC Printing.
5. Cramsession.com. 2007. Building a defense in depth toolkit. Retrieved March 1, 2007, from http://www.cramsession.com/articles/get-article.asp?aid=1105
6. Viega, J., and G. McGraw. 2002. *Building secure software: How to avoid security problems the right way.* Boston, MA: Addison-Wesley.
7. Garcia, Op. cit., 35.
8. FEMA. Are you ready guide. Website: http://www.fema.gov/areyouready/
9. http://sema.dps.mo.gov/04%20Business%20plan.pdf
10. http://www.fire-extinguisher101.com/

Chapter 3

Security Design and Planning

Paul R. Baker

Security Planning and Evaluation

When planning for security operations, a security professional must take into account people, procedures, and technology. All of these areas are intertwined and must all be considered and accounted for if a successful security plan and continuity of operations plan (COOP) is to be established. This plan must also be connected to an incident threshold to determine the appropriate level of response toward events.

People are the most important consideration for any security plan. People are not considered expendable assets. Yes, people can be replaced with new personnel, but they are still an asset that must be protected to the highest extent possible. Personnel within an organization have specific functions depending on the department in which they work and the expertise they possess. While planning for security measures, professionals must analyze the movements and requirements of employees, vendors, workers, and visitors[1] (Figure 3.1).

Each of these primary groups of people must be accounted for in a security plan and need to be planned for during emergency preparations. During normal operations the people of an organization may be used as a form of informal surveillance and monitoring as they can normally recognize irregularities and problems as well as notify appropriate responders. In addition to the majority of workers in an organization, the people of the security team provide further protection, serving at all levels of the security spectrum. The

Figure 3.1 It is important to understand employee, vendor, and visitor flow and movement within a facility for safety and security implications.

size of the security force will vary depending on the type of organization, size of the facilities, and type of asset to be protected. While workers may only be capable of identifying a threat, the security force should have the capability of further delay and proper response.

The people within an organization provide both a model to be planned around as well as a tool to be utilized in the preparation of security procedures and emergency preparedness. While technology and planning are able to augment the workload placed on individuals and possibly defer the cost of security, people are an organization's first line of "thinking" defense and are the only portion that is able, given proper information and planning, to respond in a method in accordance with the organization's goals.

A successful plan takes into consideration which department needs to be functional and back in service in the shortest amount of time possible. It is important to understand the pattern of movement of a building's occupants. This will help to ensure the procedures outlined in the plan and take into account where the highest concentration of personnel may be located. With the examination of people within an organization, it is important to determine the type of security personnel an organization will use. Contract or proprietary personnel can be utilized. The most effective and efficient methods in association with the locations to deploy the security personnel assets must also be determined.

Procedures are plans specifying the movements and contacts that need to be made after implementation of the security plan. Theses plans normally include evacuation procedures, assembly points after evacuation, and, most importantly, recovery plans to establish normal operating procedures within the shortest amount of time possible. These plans are established after a risk assessment is conducted to identify the most likely threats to an organization. The identification of these threats will also identify their capabilities and what adversarial capabilities need to be defended against. These plans also must include what to do in case of natural disasters as well. These procedures are later used by security personnel to help the organization recover from untoward events occurring to the organization.[2]

Research has shown that an organization that has planned in advance for adverse conditions acting on its operations will recover more quickly and will function more predictably in a given emergency situation. With this in mind, it is evident that development of a planned response to emergency situations is critical to organizational success. While it is impossible to plan for every possible contingency, it is important that an organization have a basic strategy for dealing with unplanned circumstances. This strategic plan, like security, must incorporate people and technology in appropriate ways.

Though it is not possible to plan for every contingency, it is important to understand natural advantages given the organization by member assets. Many people react to stressful events irrationally; however, there are always those that will "rise above the chaos and respond with calm and clear thinking,"[1] which may be utilized to an organization's benefit by providing training and emergency preparedness plans and resources. Also, much of the technology already in place, such as closed-circuit TV (CCTV) or passive and active access control initiatives, provide both current information such as the number of people in a facility and their location, as well as future tools for after-action reporting and evaluation[3] (Figure 3.2).

Possibly the most critical element to the implementation of an organization's plan is communication. Many recent events, including earthquakes in California and Hurricane Katrina in New Orleans have demonstrated the effects of positive communications as well as breakdowns in communications between executive level leadership and tactical forces. The basic demonstration in either case is the same: Communication is the link needed to

Figure 3.2 A hooded outdoor PTZ (pan-tilt-zoom) camera.

execute any plan—whether improvised or foreseen—and the presence of a good plan combined with proper training provides a faster and more thorough solution to most emergency situations.

Technology demands are dependent upon what is appropriate for the mission of your facility. It is important to have only the most appropriate security equipment for your facility (i.e., x-ray machines for parcels and bags coming into a facility) if that is deemed necessary due to the functions occurring within the facility. Technology provides the needed cost deferent in order to provide proper security to most facilities as well as useful tools to improve the performance of a security or disaster preparedness initiative. At all stages of security as well as emergency preparedness, technology is able to enhance the effectiveness and possibly defer the cost.[3]

CCTV systems linked to intrusion detection devices can significantly reduce the number of security personnel needed to patrol a facility. In addition, the same technology provides a unique opportunity to assess a threat in order to provide a more useful and appropriate response. Technology is further able to aid in the screening and access of individuals allowed to enter a facility.[1] Card readers, smart card systems, and other entry monitors are simple but effective means that substitute for a manned entrance (Figure 3.3a and b).

While these systems may not be appropriate for all organizations, when they can be used they are a cost-effective means of entrance security and surveillance that provide useful information to both security professionals and emergency first responders. These systems provide both detection and response capabilities; however, it is important to remember the deterrent possibilities of technology. The deterrence presented by technology

Figure 3.3 (a) Key pad on a security door and (b) card reader.

is dependent almost entirely on the level of threat against an organization. While a vandal or other thief may be easily deterred by the possibility of detection and resolution presented by CCTVs mounted on building walls or in parking lots, a determined criminal with sufficient reason and resources may not be sufficiently deterred by such simple technological implements.[3] While deterrence is not in itself a viable security solution, implementation of technology in the pursuit of security objectives and emergency preparation does have this added benefit and can contribute to a lower cost of doing business as well as a more secure environment for work.

Another consideration for the plan is the establishment of an incident threshold. The security manager must have a predetermined threshold of what can and cannot be handled from within the security department. If an incident is determined to be small enough to be handled from within the organization without executing the security plan in full, then that will keep costs down. The full implementation of the security plan will cost money—not only in terms of response by outside entities such as police and firefighters. It will also cost money in terms of lost productivity by the employees. If they cannot do their jobs, then that will cost money.[2]

In accomplishing the goals of detection, delay, and response in a total security solution, an organization creates the baseline needed to plan for emergency situations and prepare its people for proper response. The actions taken when developing security solutions—including installation of CCTV, which provides surveillance, deterrence, assessment opportunity, and archiving; screening; access control and monitoring; emergency response time evaluation; etc.—develop a solid foundation on which to build proper response levels for varying levels of threat.[3] It is the proper mixture of people, planning, and technology that will lead, as demonstrated throughout recent history, to a faster and more effective solution to security threats as well as emergency situations.

All of these areas must be considered when establishing security operations and continuity of operations. All aspects must be examined and considered when creating a document to establish emergency procedures. These plans must also be updated at established intervals to ensure contacts are updated within the local communities.

Security Design Concepts

When considering elements for a security program, ideas and thoughts need to start early in the planning process. Integration of technology, personnel, and policy/procedures during this phase will allow the planners to grasp the concepts and design attributes to ease in their implementation. Effective plans will ensure that emergency contingencies are covered, reducing some risk and providing overall enhanced security for the organization.

Design includes architecture, engineering, landscaping, and site planning. The first line of defense is the site perimeter. Building placement is important relative to the ability to have access control and natural surveillance. Building orientation can have an impact on the spatial relationship to the site, its orientation relative to the sun, and its height in volume relative to the site. The placement and amount of open space are important to site security, specifically as it applies to the ability to observe the perimeter and detect intruders, vehicles, and hidden contraband. Another benefit of open space is the standoff distance from the blast energy of an explosion.

Designers should understand that the impact of vehicular circulation patterns could make a facility or site more vulnerable if security implications are not carefully considered. The designer can propose a roadway system to minimize a vehicle velocity, thus using the roadway itself as a protective measure.[4] Straight-line approaches to building should not be used, because these give cars and trucks the opportunity for gathering speed and momentum to crash into the building.

Raised entrances, lower landscapes, dirt berms, strategically placed decorative boulders, bollards, fencing, and curved driveways can render vehicular attacks difficult or impossible. Architectural landscape elements like raised planter beds, park benches, lampposts, serpentine or curved roads, traffic-calming devices, site lighting, and trash receptacles serving as a vehicle barriers prevent a direct line of approach to a direct line of sight to the building.[5]

Buildings with layers of buffer zones are much less appealing targets than those without them. Placing visitor parking in an area not directly adjacent to the building can also be a deterrent insofar as terrorist vehicles cannot be parked for detonation purposes in an effort to destroy or damage the building. Natural surveillance, or the ability to see with an unobstructed view of who is entering the property and whether it is for a legitimate use, is an important part of the safety plan. The best placement of the footprint of a freestanding building is one that has all facades set back from the street.

Beyond environmental design, other relatively low-cost precautions can be used as bases to protect vulnerable utilities by locking manholes in the street in securing areas such as electrical rooms, fan rooms, mechanical rooms, and telecommunications space inside the building with locks and alarms.[5]

Other measures include developing operational policies and procedures including the increased use of security personnel, motion sensors, cameras, and CCTV as part of the more comprehensive security plan. Biometrics, proximity card readers, and other forms of electronic access controls, along with defenses integrated into heating, ventilation, air conditioning HVAC, and other systems, can be used. To counter a chemical attack, defense design experts recommend installing air intake above the grade on the roof at least three or four stories aboveground.[5]

Other built-in defenses for existing buildings could include air-detection systems and carbon filters to protect against chemical release in high-efficiency particulate air filters used in conjunction with UV3 light wands to achieve a high capture rate of spores or other biological agents. In existing buildings, where renovating the entire HVAC system is likely to be cost prohibitive, an alternative is to put vulnerable areas such as the lobby, mailroom, and receiving dock on a separate HVAC zone. Then, if there is a release in the lobby, it is not transmitted throughout the entire building. The design also can positively pressurize the building lobby or even the whole building; in the event that there is an external release, it is kept outside those areas by the pressure.

Pressurizing a building and locking it down can protect against external release. The goal in defending against an internal release is to isolate that contaminated area and purge it as quickly as possible. The process of protecting people and property against terrorism is much easier to implement in new buildings. In a new building the cost to isolate the lobby, the loading dock, and mailroom areas is minimal, and a building owner should incorporate that level of the defense.

By considering what technologies are used in the design of the building, the security professional can determine cost/benefit of certain types over others. This is an important factor for stakeholders, who are already budget conscious and are trying to save wherever possible. Some technologies may be self-sufficient, such a REX (ready to exit) being used to release the door lock as a person approaches the door (Figure 3.4).

But others may require a human element to be effective, such as an alarm key pad requiring an individual to arm the room when he or she leaves. Technologies can also be concealed within the framework of the building, going unnoticed by its occupants—for example, a motion detector placed

Figure 3.4 Ready to exit (REX) device above an exit sign.

high on the wall in the corner of the room. With the use of these technologies, procedures have to be developed for security personnel to follow. The best technology in the world is useless if an operator cannot analyze an alarm to determine a false, nuisance, or threat situation. When technology is properly integrated into the overall security program, it allows the free flow of information for the security force.

People are the single biggest asset for any corporation. Without its workforce, any company loses it ability to complete the necessary tasks to create success. This also affects its market share, profitability, and future competitiveness. Developing security procedures to include the human element allows the designer to foster a safe/secure work environment. Along those same lines, the type and number of response personnel has to be addressed within the plan. This part of the plan should address at what point the force should be recalled, what level of intervention will be used, and how they will coordinate with traditional law enforcement. Exercise plans and coordination with local authorities are essential to support these plans. Analysis of threat, risk assessments, and asset management all play an important part in security plan development.

The security professional will be required to define the threat. This involves considering factors about potential adversaries: class of adversary, adversary's capabilities, and range of adversary's tactics. Next, the security professional needs to identify targets: "what to protect against whom."

This information allows for the security professional to begin the design stage of the protection system. What will be the necessary requirements for protecting the facility? Will it need fences, vaults, sensors, procedures, communication devices, and protective force personnel?

Once a security system has been designed, it must be analyzed and evaluated to ensure it meets the organization's security objectives. Due to the complexity of protection systems, an evaluation usually requires modeling techniques. If any security breaches are found, the initial security system must be redesigned in order to correct the vulnerabilities. This will lead to a fully integrated, fortified, system-tested facility (Figure 3.5).

Developing specific plans and procedures is the single most important aspect of security plan development. By telling security elements exactly where to go, what to look for, and what level of force is authorized, deviation is limited and accountability is maintained. Planning for the unexpected is also necessary. This gives the responding force the ability to think independently of management and to intervene as necessary to be successful. Careful decisions must be addressed and communication

Figure 3.5 No facility is an island. Integrated physical security, security personnel, procedures, and multifaceted layers of protection all contribute collectively to a safe and secure environment.

is paramount in these instances. The operations control element is the base for all activity. It may be located within the security workspaces or corporate headquarters or farmed out to a support company. Even natural disasters and internal building problems have to be addressed. These response plans may be considered tedious in nature, but when a fire breaks out in an office or common area, everyone needs to know exactly what to do.

The workforce also has to be cognizant of the policies that provide a secure workplace. All levels of the workforce—labor, management, and executives—must review and understand these procedures. By knowing exactly what is expected of them, the workforce is able to facilitate evacuation, if necessary, or deter a threat by augmenting physical security assets in place, such as securing door/window locks, phones, or automation equipment.

Security operations are always a challenge. From the stakeholders' standpoint, they may be seen as money wasted or otherwise better used elsewhere. From the management and labor point of view, policies may be considered a nuisance as they may interfere with getting the work done on time. It is important for the security professional to be integrated early in the development of building operations. This ensures that technology, policy, and people are considered—not if a security situation arises, but when the worst happens.

Security Technologies

The ever-changing world of technology is the constant. The security professional is not immune to its effects. Within any organization, cost is the driving factor on equipment purchases. The plan is to have a comprehensive approach toward the selection of equipment needed to perform the security function. The problem is the amount of security equipment out there and the time it takes to review and analyze. Most of us know what we want it to do but do not have a understanding of the inner workings or technical capacities. We typically rely on vendors, and if you are fortunate, you will have an in-house security engineer you can advise on the competencies of the product.

Equipment needs to be identified with a life-cycle replacement. If a DVR (digital video recorder) is on the books for depreciation over a 5-year period, then it should be reasonable that a replacement should be budgeted. However, this is the quandary: The system that is in place needs to have the flexibility of growth. Newer and better products need to be installed into an existing platform. The question is whether this can be done with a software or firmware upgrade. In the case of a DVR, is it more reasonable to go to the next generation—NVRs (network video recorders)? This is where your worth as a security professional will show its true mettle.

The security industry is huge. ASIS (American Society for Industrial Security) International has a 4-day security conference and exhibit every year. The ASIS website indicates that the exhibits highlight

> ...cutting-edge solutions to integrated tools designed to improve efficiency; ASIS is your best one-stop opportunity to see, test, and compare everything you need to meet your security challenges in the coming year. Some 850 exhibiting companies will showcase the most innovative technologies, products, and services on the market today. From access control to weapons detection and everything in between, ASIS exhibits ha[ve] it all.

This is where your education on product lines and security equipment will be under one roof. Vendors, experts, and colleagues will all be at one location, which will allow you to shop until you drop with security questions and concerns. All will want to demonstrate their capabilities and show that they are the best and brightest in the business. This is truly a three-ring security circus, but you do not want to miss it.

From the standpoint of doing risk assessments of the facility and to validating your physical protection security system, several security models have been developed to analyze physical protection systems.

EASI Model

The EASI model is a calculation tool that is simple to use and quantitatively illustrates the effect of changing physical protection parameters along a specific path. It uses detection, delay, response, and communication values to compute the probability of interruption. However, it can only analyze one adversary path or scenario at a time. Even so, it is able to perform sensitivity analyses and analyze physical protection system interactions and time trade-offs along that path.

For unauthorized intrusions by an adversary, a response team needs to be notified in plenty of time to defeat the attack. The input for the model requires (1) detection and communication inputs as probabilities that the total function will be successful and (2) delay and response inputs as mean times and standard deviations for each element. The output will be the probability of interruption or the probability of intercepting the adversary before any theft or sabotage occurs. After obtaining the output, any part of the input data can be changed to determine the effect on the output. However, since EASI is a path-level model, it may be necessary to use another model to observe all possible paths to determine which are the most vulnerable.

Adversary Sequence Diagram

In a typical facility there can be several options for an adversary to defeat the levels of security protection. An adversary wishing to penetrate a locked building can defeat doors, windows, or walls or enter through the roof. Because there are several entry points and paths into a facility, an EASI needs some systematic method of recoding these paths. The method used is the adversary sequence diagram (ASD).

The adversary sequence diagram is a graphical representation of the physical protection system at a facility. It identifies paths that adversaries can follow to accomplish unauthorized entry, sabotage, or theft. The most vulnerable path can be determined and used to measure the effectiveness of the entire physical protection system. There are three steps in developing an adversary sequence diagram for a specific site:

1. Model the facility by separating it into adjacent physical areas.
2. Define protection layers between the adjacent areas. Each protection layer includes one or more elements, which are the basic building blocks of a physical protection system. Examples are doors, fences, surfaces, and security portals.
3. Draw path segments between the areas through the protection elements. Both entry and exit paths can be modeled. Paths that appear to be the weakest can be entered into the EASI model to be calculated and compared.

Working with Architects

Architects need to be reminded that the design of an effective physical protection system includes the organization's security objectives. This will require the evaluation of the design and probably a redesign or refinement of the architectural drawings. To develop the objectives, the physical protection professional must have gathered information about facility operations and conditions, such as a comprehensive description of the facility, operating states, and physical protection requirements. Knowing the security needs will provide the architect with the necessary design criteria when initiating the project.

Typically, architects send CAD (computer-aided drawing) for review by the facilities manager, program manager, and security. Everyone who is a stakeholder will be asked to review and comment on the drawing prior to construction. This is the time for a security professional to assess and question details of the construction before actual work begins. It is easier to change a drawing than to do a change order during construction.

Working with Contractors

If you do not have the luxury of having your own alarm technician on staff, you may require the services of contractors. This is what the majority of the industry relies on and a multitude of qualified security technicians is available. The problem is in finding the right company to service your security needs. There are basic rules and ideas in how to deal with contractors.

Have Basic Knowledge of Your World

Contractors will sell you anything. Their business is to sell you their product and services. As a security professional, you need to have a basic understanding of equipment requirements and applications. This can be accomplished by working with your IT group to provide you with expertise when sitting down with a contractor. The idea is not to try this on your own; regardless of your expertise, it is always better to have more eyes on a project then less. This book provides the foundation for a security professional. There are also several books available on a wide range of security subjects to assist. ASIS has a website and a monthly magazine dealing with all security subjects and equipment critiques. The idea is to be a well-educated consumer dealing with security applications and equipment. Your company is entrusting you to provide a cost-efficient and functional security operation.

Know What You Want

This is critical. Too many people sign up with a contractor in order to proceed with a certain job, but in the middle of the job, they realize that they wanted something else. Maybe they wanted to add a card reader; this requires wiring and having enough room on the panel to add the reader. In the middle of the installation, they decide to add another reader or have it moved to another point. This is a big deal and can wreak havoc in a contractor's plans, perhaps causing him or her to have to undo some of what has been done or to spend more time than what was budgeted on your project. Change orders can end up costing you a lot more than if you had been exact about what you had wanted at the outset.

Just as with any project, have an idea what is needed and ask your contractor questions. For example, consider the things that will need to done for a door: type of door (metal or wood), thickness of door, door frame, locking hardware for door, requirements for egress, color of door hardware, door closer required. Here is where a good relationship is very helpful when developing a requirements list for any project.

Develop a Relationship

It is easy to forget that security contractors have other clients. Some contractors have a fleet of service technicians and you may get a different technician every time you call for service. Others may be a mom-and-pop outfit that provides individual attention, but may not have the resources to respond to your facility at a moment's notice.

Find a contractor you really like—one you feel you can trust and with whom you will be comfortable communicating. You should be able to ask questions and get answers you understand. Ask around. Do not just rely on the Yellow Pages in your search. Talk with colleagues and see what companies they are using and what equipment they deploy. See who has had good experiences with their contractors. Do the due diligence and find contractors who are trusted and competent.

Budget

You are on a budget, right? Well, make sure your contractor is fully aware of it. Expect some variation between the estimate you get and the final cost. You might even want to plan to spend 10% more, in total, on the job. Why? Well, things happen. Over the course of a few months, the price of wiring may skyrocket. The cost can also rise if you keep adding to the scope of the work. Once you, as a security professional, have the estimate, then funding can be approved for the project. Understand that there is a process for doing security projects.

Typically, if you already have an established relationship with a contractor, then the first step is to obtain a written estimate for the work. If you are doing government work, then it will require that you obtain three estimates unless you can articulate that it must be sole sourced. In either case, the estimate is the initial step.

One good way to discuss your budget with a trusted contractor is to say something like, "I have a budget to spend no more than X dollars on this job, but if need be, we can spend an extra X dollars—and absolutely no more than that." This can help the contractor to decide where he or she can upgrade materials and where he or she must be as frugal as possible.

Construction Review

The review process begins with the initial estimate. Once the contractor has the concept of what you want, then this will be processed with an equipment list. This equipment should be inserted into as-built drawings that will identify where each piece of equipment will be installed. This is your first assessment to see if the contractor has the same concept as you have been trying to propose. Having the concept on a drawing is just that—on a drawing. This will require you, as a security professional, to plan visits to the site to verify that work is being done correctly.

If the contractor has a project manager, then this is the person you need to be in contact with and have along for the visits. Contractors stay within the scope of the project and do not waver outside the scope without the business owner's permission. If, during your walk-through, there are discrepancies or things do not look the way you envisioned, then this is the time to discuss this with the project manager. If needed, a change order should be initiated. The number one rule of construction is to expect change.

There needs to be a final walk-through with the contractor in order for the security representative to sign off on the project. This can be accomplished with a test and acceptance process. Once this is completed, continue to verify with the project manager every aspect of the project from what was installed—matching this up with the equipment list. Make sure that all components that have been installed work properly and to the specifications that were agreed upon; this can be accomplished during the test and acceptance portion of the final punch list. If, as a security professional, you are satisfied, then it is now up to you to sign off on the project and send in the invoices for payment. Typically, the cost of the project is not paid by the company until all work is done to the satisfaction of the security representative.

References

1. Grassie, P., and B. E. Behrooz. 2004. Facility management: Building security access control measures. AIA Best Practices. January 2004.
2. The American Institute of Architects. 2004. *Security planning and design: A guide for architects and building design professionals*. Hoboken, NJ: Wiley Publishing.
3. Garcia, M. L. 2006. *Vulnerability assessment of physical protection systems*. Burlington, MA: Elsevier Butterworth-Heinemann.
4. FEMA. 2003. 426 2-11. http://www.fema.gov
5. Atlas, R. 2008. *21st Century security and CPTED*. Boca Raton, FL: CRC Press.

Chapter 4

Security Construction Projects

Paul R. Baker

New Construction

If you are fortunate to get in on the ground floor of a construction project, make sure that your security voice is heard. Do not allow architects, engineers, and facility personnel to dominate the process. Security is an integral part of the process and it can also be a cost savings.

There is *no* requirement that you as a security professional have extensive knowledge of systems engineering. But you will need the support of several specialists to make the project smooth and seamless. Regardless of whether it is a new construction project or a retrofit project, it is important to have a knowledgeable technical person involved in order to provide a comprehensive scope for the project. Depending on the size of your company, this person could be within Security, Facilities, or the Information Security groups. There is also the possibility of assistance of an outside contractor or a technical team, depending on the size and complexity of the project.

Finding a qualified contractor to assist in any project is, in itself, a gamble. A contractor will tell you anything because his motivation is to find and keep that "cash cow." This requires due diligence on your part—knowing who you are dealing with. Someone that you have worked with in the past and feel comfortable that you will get straight and honest answers is what you want in a contractor. This continually comes back to the main message: "Having a qualified security team is essential; you cannot do it alone."

It is best to have a network of security professionals that you can reach out to and ask the question, "Who are you using and would you recommend them?" Depending on the size of the contract, your company may require that you have several bids and will require an RFP (request for proposals) or RFQ (request for quotes). You will need the assistance of your Purchasing or Finance group to assist in the writing of your proposal. This exercise will allow you to write specific requirements that the contractor will need to answer. You have the luxury to have the vendors come in one at a time and do a live presentation, explaining what services they will provide and how they plan on accomplishing the task. This allows you to get a feel for each vendor and also a reference list that you can use to contact other security professionals and ascertain their feelings on the vendor's performance and work product.

> My company was in need to recompete for a security guard contract. I had a copy of the last RFP the company had used, and I also went online and reviewed several other guard RFPs. I put together a new and comprehensive RFP and sent it out to several guard companies. Each one responded and sent me a proposal and pricing list. I started with seven companies with the idea to cut it to three, who would then come in and give a face-to-face presentation. As I began reading the first proposal, I noticed right off the bat that it was a cut-and-paste job. The cover sheet indicated that this company was presenting to a completely different company. No attention to detail—if they could not even send in a proposal with the correct company's name, how effective would they be as a guard company?

Once the security team is together and all parties are reviewing the project, this is the time to develop the security envelope for the new facility. The review needs to start at the edge of the property line and work inward. How can crime prevention through environmental design (CPTED) be used? (This will be discussed later.) What materials and products need to be installed? This list will include CCTV, access control, intrusion detection, lighting, guard services, and locking hardware—all responsibilities of the security team.

Another issue that will require the expertise of the security group concerns the risk of and vulnerability to bombs and the requirement of blast protection. We are now in the age of terrorist bombing and the approach needed is to set the stage for a secure facility that can withstand a blast and have the building materials that will protect the inhabitants.

As a security professional considers the measures for explosive blast threats, the primary strategy and most commonsense approach is to keep explosive devices as far away from the building as possible (maximize the standoff distance).

Standoff distance is the closest a threat can come to an asset or building. For blast concerns, standoff distance is measured as the distance from a building's exterior to the nearest vehicle access, such as a road, parking lot, service yard, or an area of unrestricted movement across the landscape.[1]

This is the easiest and least costly way to achieve the desired level of protection for the facility and the personnel. In cases where standoff distance is not available to protect the building, hardening of the building's structure may be required, as well as design to prevent progressive collapse. In addition, a designer should try to minimize hazardous flying debris during an explosive event because a high number of injuries can result from flying glass fragments and debris from walls, ceilings, and nonstructural features.

A structure to consider is the "warehouse" style of structure. This style will provide advantages to the security envelope by the mere concept of a bunker style structure. To further the protection, this structure allows for the distribution of people, assets, and operations across a wider area. Even with a low style of structure, the shape of the building is also a factor. The U- or L-shaped building will trap shock waves, which may exacerbate the effects of the explosive blast. It is recommended that re-entrant corners be avoided. In general, convex rather than concave shapes are preferred for the exterior of the building. A circular building acts to reduce the air blast pressures because the angle of incidence to the shock wave increases more rapidly than in a rectangular building.

For structures that are considered "high rises," there are ways to reduce the building's vulnerability to attack. This can be done by placing the ground floor elevation of a building at 4 feet above grade to prevent vehicle ramming. Avoid eaves and overhangs because they can be points of high local pressure during blasts. Avoid exposed structural elements (columns) on the exterior of the facility.

With either design, all facilities typically have windows. The commonsense approach is that windows are the weakest point and, when they are blown in, will have the potential to cause greater injury from broken shards of glass. The simplest way to protect this portion of the building is to install laminate glass or use security window film over existing glass.

Laminate glass is two or more pieces of glass bonded by a polyvinyl butyral plastic interlayer. Compared to conventional glass, laminated glass can provide increased resistance to windblown debris and seismic and explosive forces.

Security window film can be applied to both single panes and many types of insulating glass. Proper application of appropriate film to insulating glass does not impact the integrity of an insulating glass sealant or generate thermal stress to glass from uneven heat absorption. Because security window film has the ability to stretch without tearing, it can absorb a significant degree of the shock wave of an explosion.

Initial Point of Access

With a new construction design there are many other areas of concern outside the physical building itself. As a security professional, you will have to address the questions: Where does the point of security start? Is it at the end of the parking lot or the fence line? Does it traverse out beyond the property line and the community in order to maintain a visual watch for potential threats? The farther one can observe allows for more time to react. According to Philip Purpura, "The design of roads and parking lots influences the security at the building site."[2]

Traffic monitoring can be accomplished in several ways to include the design of the roadway to monitoring with CCTV. The most controlled way is to have an entry control point or guard building serving as the designated point of entry for site access (Figure 4.1). It provides a point for implementation of desired/required levels of screening and access control. The objective of the entry control point is to prevent unauthorized access to the facility while maximizing the rate of authorized access by foot or vehicle. Designs should be flexible to allow implementation of increased security controls when organizations are placed in high alert and easing of controls at lower threat levels.

Roadway Design

There is not a facility anywhere that does not have roadways and vehicular traffic. Employees, visitors, and deliveries have requirements to bring their

Figure 4.1 A guard booth.

vehicles close to the facility. The concept of designing streets and roadways as a way to curtail unauthorized access or prevent sabotage and structural damage to the facility is never thought of as the first line of defense. Streets are generally designed to minimize travel time and maximize safety, with the end result typically being a straight path between two or more end points. Although a straight line may be the most efficient course, designers should consider a roadway system to minimize vehicle velocity, thus using the roadway itself as a protective measure. This is accomplished through the use of several strategies.

Straight-line or perpendicular approaches to the facility should not be used because this gives a vehicle the opportunity to gather the speed necessary to ram and penetrate buildings. This can also occur by accident when a gas pedal sticks and the driver panics. Instead, approaches should be parallel to the perimeter of the building, with natural earthen berms, high curbs, trees, or other measures used to prevent vehicles from leaving the roadway. Existing streets can be retrofitted with speed bumps, barriers, swing gates, or other measures to force vehicles to travel at a slow pace as they approach the facility. The front of the building can be equipped with decorative bollards to keep vehicles from running into the facility (Figure 4.2a and b). In high-security areas, "delta barriers" are used to control vehicle traffic into secured areas (Figure 4.3).

Figure 4.2 (a and b) Bollards can come in various shapes, sizes, and designs to match a facility while still protecting against possible vehicle threats. (Photo in Figure 4.2b by Daniel J. Benny.)

Parking

Parking restrictions can help to keep potential threats away from a facility. In downtown settings, underground parking is often necessary and sometimes difficult to control. Mitigating the risks associated with parking requires creative design measures, including parking restrictions, perimeter buffer zones, barriers, structural hardening, and other architectural and engineering solutions.

Figure 4.3 Delta barriers in a parking garage. Delta barriers are used for high-security applications. (Photo Courtesy Randy Atlas from *21st Century Security and CPTED: Designing for Critical Infrastructure Protection and Crime Prevention,* 2nd ed., CRC Press, Boca Raton, FL.)

Visitor parking should be located in the front of the facility in full view of the reception/guard desk. There should be limited space availability for guests and visitors. Some facilities hold large meetings and, in these situations, designated areas should be cordoned off specifically for those events.

Facilities should also encompass a clear zone area (Figure 4.4). This can be achieved with perimeter barriers that cannot be compromised by vehicular ramming. A continuous line of security should be installed along the

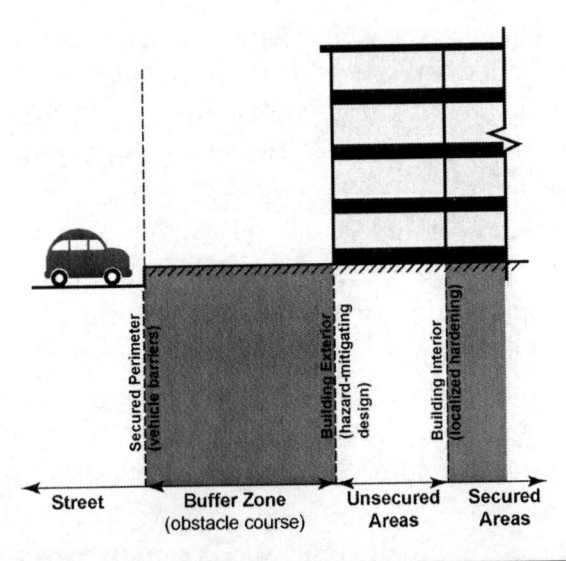

Figure 4.4 Clear zone area. (From FEMA 427. 2003. Primer for design of commercial buildings to mitigate terrorist attacks.)

perimeter of the site to protect it from unscreened vehicles and to keep all vehicles as far away from the facility as possible.

According to FEMA,[3] the following critical building components should be located away from main entrances, vehicle circulation, parking, and maintenance areas. If this is not possible, harden as appropriate:

- Emergency generator, including fuel systems, day tank, fire sprinkler, and water supply
- Normal fuel storage
- Telephone distribution and main switchgear
- Fire pumps
- Building control centers
- Uninterrupted power supply (UPS) systems controlling critical functions
- Main refrigeration systems if critical to building operation
- Elevator machinery and controls
- Shafts for stairs, elevators, and utilities
- Critical distribution feeders for emergency power

Parking Garages

If the building has an underground parking facility (Figure 4.5), the security professional will have to understand that there will be two primary safety threats: crime and vehicles hitting pedestrians.

Start by controlling access into the garage area. This can be accomplished with automatic controlled gate arms that will be activated by using a transponder or a badge reader. The badge reader can be incorporated into the overall access control system and the same badge that will be used within the facility can be used to enter the garage (see Figure 4.6).

For additional security, industry advancements have led to the development of quick-access roll-up doors, which have dramatically reduced the time it takes for the doors to open and close. With this system, an access card is inserted and the door opens within 7 seconds and then immediately closes. This system is a vast improvement over traditional roll-up doors that take anywhere from 20 to 25 seconds to come back down again, allowing time for unauthorized individuals to follow cars into the building on foot.

The use of signage can direct vehicles and pedestrians to the exit points from or the entrance to the facility. The only entry points into the garage should be for vehicles. Inside the garage, signage should be easily

Figure 4.5 A typical underground parking facility.

identifiable to direct pedestrians toward elevators and stairs. CCTV cameras should be used for monitoring events and emergency call boxes should be placed throughout the garage. Generally, when the parking floor is only 200 to 250 feet long, only one camera needs to be used at the end of each aisle. For floors from 250 to 400 feet long, a second camera can be placed midway down the aisle so that the cameras do not have to be manipulated for telephoto or tilt zoom. Cameras should also be positioned to capture activity in and around elevator lobbies and in stairwells.

Figure 4.6 A gate with automatic controlled arm.

Installing bright lights is one of the most effective deterrents to both accidents and attacks. Lighting levels of at least 10 to 12 foot candles over parked cars and 15 to 20 foot candles in walking and driving aisles are recommended.

It is also advisable to install high lighting levels to illuminate the exterior of the parking facility, particularly in areas that experience high pedestrian traffic. As a rule, exterior lights should be placed approximately 12 feet above ground, and they should point downward to illuminate wide areas along the ground. Another method for increasing visibility is to paint the walls of the structure white to reflect light. Lighting fixtures should also be strategically placed to bounce light off the walls and reduce dark corners where criminals could hide.

With garages under the facility, elevators or walk-ups should all empty into the lobby, outside the controlled space. Having all employees and visitors pass through the controlled receptionist area will maintain the integrity of the facility. In this way the elevators going into the core of the building will only be accessible from the lobby and not from the garage levels.

It is always best to install emergency communications systems (e.g., intercom, telephones, etc.) at readily identified, well-lighted, CCTV-monitored locations to permit direct contact with security personnel.

Open Area Parking

In a rural or industrial area, open surface parking areas are common (Figure 4.7). The way to provide measured security is with initial parking access control. Only allow employees access into specific segregated parking areas. Again, this can be accomplished with the use of automatic gate arms that will be activated by using a transponder or a badge reader. Both devices will allow access into the parking area. It is best to have an entry-/exit-only design, which will maintain a one-way circulation within a parking lot to facilitate monitoring of potential aggressors. To maintain a secured parking area, it is best to locate parking within view of occupied buildings while maintaining standoff zones and to prohibit parking within the standoff zone. The parking lot could also utilize control arms for intended employee vehicles only.

Figure 4.7 In addition to sharing safety and security concerns similar to those for garaged parking, outdoor parking areas require other unique security considerations.

Loading Docks

All facilities will require deliveries, and loading docks and service access areas are typically desired to be kept as invisible as possible (Figure 4.8). For this reason, special attention should be devoted to these service areas in order to avoid intruders. Who controls the deliveries to the facility? Is it a security guard making sure that all deliveries are recorded? Does security work in conjunction with a warehouse, mail room, or shipping and receiving group to coordinate deliveries on the loading dock platform? The operation should control the overhead doors to the loading dock just as it would protect perimeter doors. When these doors go up, a physical security presence needs to be in place. From the standpoint of a perpetrator, this would be the idea location to gain entry. Daily deliveries and pickups are made all day and workmen and contractors move supplies in and out of the facility.

Some organizations design their loading docks using an interior door control. In this scenario the loading dock doors are controlled by a card reader, which would only allow authorized persons to operate the roll-up doors. When the overhead doors go up, the interior door that leads into the facility is controlled to lock and cannot be opened until the loading

Figure 4.8 Loading docks offer unique security issues, given the transfer of goods and services, interaction of employees and vendors, and the potential for intruders to access the building interior undetected if it is not properly monitored.

dock doors are shut. This keeps a secure perimeter even when deliveries are being made.

Proper loading dock security starts with comprehensive policies and procedures for shipping and receiving operations. Here are some examples of simple steps to improve operations that many organizations fail to implement:[4]

- Secure loading dock overhead doors and entry doors
- Implement personnel screening processes to ensure that only authorized personnel access the loading dock areas
- Institute detailed logging of all items entering or leaving the facility
- Implement security awareness training programs for nonsecurity personnel in order to communicate that security is everyone's responsibility

Signage

Signs are an important element of any facility layout. Signage is necessary to direct visitors, deliveries, and employees to their respective parking areas. Signs are also meant to keep intruders out of restricted areas and make notification that there is "no trespassing" (Figure 4.9). Signs should be located at the street entry of the facility and should explain current entry procedures for drivers and pedestrians. They should give traffic regulatory and directional details that control traffic flow and direct vehicles to specific appropriate points. Post easily understandable signs to minimize accidental entry by unauthorized visitors into critical areas. In areas where English is one of two or more languages

Figure 4.9 Clear signage is necessary to direct visitors, deliveries, and employees and to deter individuals from restricted areas.

commonly spoken, warning signs must contain the other language(s) in addition to English. The signs should be posted at intervals of no more than 100 feet and should not be mounted on fences, posts, or light poles.

Retrofitting

In building a security design, technologies can be built in or retrofitted into the existing structure to perform a variety of functions such as access control, surveillance, personnel screening, intrusion detection, and fortification. A retrofit project requires very careful planning and a seamless transition from an existing system to an upgraded system without causing stress on the facility and personnel during the process.

A security retrofit project can be at almost any level of the security system, any functional security area (CCTV, alarms, access control, etc.), or the security control center (SCC). I had a situation where the SCC needed to be expanded due to the overall new building construction of the operation. The overall size of the operation was growing and it required a larger and more functional room to house the SCC.

When this type of situation arises, the first thing a security professional should do is to ask the following questions: What are the needs? Can you upgrade the existing products? Do you need to install a new system? Will the systems integrate with each other?

Robert Pearson, a security engineer, provides an example of problems that can arise during a retrofit:

> During the conversion of one alarm system to a new alarm system, the existing alarm system must remain operational through the retrofit effort. Continuous operation can become a major challenge—particularly in facilities that have very limited space for the security system field equipment to reside, which is usually the norm.
>
> When a new alarm field panel must be installed and there is no wall space, careful thought must be part of the retrofit project to cover the cost of dual monitoring and ensure that no sensor alarm is missed. One way to address this specific problem is to leave the old system running while the new system is being installed. In small areas, where the alarm inputs cannot be changed from one panel to another within an acceptable time frame, both old and new panels must be operational. The old panel can be carefully removed from the wall and left hanging by support cables from the ceiling so that active alarm cables and other data/power cables remain operational. The new field panel can then be installed on the wall in the place previously occupied by the old panel. Then, alarm inputs are moved from the old panel to the new panel one at a time. Both systems are monitoring for alarms and no sensor is left off-line.
>
> Special consideration needs to be made in the SCC to enable the operators to provide as proficient a job as possible with both manufacturers' alarm servers sitting side by side. The more alarm panels that are in a tight space, the more likely it is that there will need to be some alarm field panels suspended from the ceiling or from a cable tray.[5]

When it comes to new construction or a retrofit design, remember that it is your reputation and career that are on the line. "The added cost of poor planning and design must be addressed at some point and it can have a negative effect on future projects and possibly the planner's career. A competent technical person or team can mitigate many of these unexpected consequences." Remember the old adage of a carpenter: "Measure twice—cut once." This is the same for any security project.

References

1. FEMA 427. 2003. Primer for design of commercial buildings to mitigate terrorist attacks.
2. Purpura, P. 2007. *Terrorism and homeland security: An introduction with applications,* 329. Burlington, MA: Butterworth-Heinemann.
3. http://www.fema.gov/areyouready/
4. Gompers, J. 2004. Security hot spot: Loading docks. SecuritySolutions.com
5. Pearson, R. 2009. Planning a major security systems retrofit, 22. Security Technology Executive, securityinfowatch.com

Chapter 5

Protection in Depth

Paul R. Baker

Protection-in-Depth Concepts

The primary goal of a physical protection program is to control access into the facility. In this concept, barriers are arraigned in layers with the level of security growing progressively higher as one comes closer to the center or the highest protective area. Defending an asset with a multiple posture can reduce the likelihood of a successful attack; if one layer of defense fails, another layer of defense will hopefully prevent the attack, and so on. This design requires the attacker to circumvent multiple defensive mechanisms to gain access to the targeted asset.

> Implementing defense in depth requires that the security practitioner understand the goals of security. Essentially, security can be distilled down to three basic elements: availability, integrity, and confidentiality. Availability addresses the fact that legitimate users require resources, which should be available to the users as needed. Integrity relates to the concept that information is whole, complete, and remains unchanged from its true state. Confidentiality can be defined as ensuring that data is available to only those individuals that have legitimate access to it.[1]

> Consider the layers of a medieval castle, which employs many redundant measures to protect its citizens and wealth. Castles were built to keep out enemies of the nobles. When an attack was expected, the drawbridge was raised, the gates were closed, and archers were stationed on the towers. In a well-planned castle, the walls not only were high, but also were arranged as much as possible so that anyone climbing the walls could be shot at from two directions.
>
> The fortress-like appearance and protective reputation of any castle is likely a deterrent factor to some who would want to lay siege to try to take control of a region. The first line of defense is the castle moat, which was used for defensive purposes and to prevent undermining of a castle. Moats were either filled with water or wooden stakes to create a difficult barrier for men and horses. If the attackers could manage to forge the moat, there was an ominous and high wall to climb over or enter through.
>
> The defenders of the castle maintained their stronghold through various means. They sat high on the walls and waited for the besieging army to come within striking distance. They constantly maintained a steady fire of rocks, torches, arrows, oil, and boiling water on the attackers at the wall. Archers were positioned all over the battlement and the towers to help with this. By maintaining this constant barrage of missiles, they could make the attackers grow weary of fighting.
>
> If the attackers breached the castle walls, they typically encountered a second interior wall that was positioned higher than the perimeter wall. If all else failed, the defenders would move back to a castle keep, which was the tower built as the most protected part of the castle (Figure 5.1).

The key to a successful defense of any protected asset is the integration of people, procedures, and equipment into a system that protects the targets from the threat. A well-designed system will provide the necessary protection in depth, while minimizing the consequences of component failures and exhibiting balanced protection.[2]

Physical protection of any facility maintains the same fundamental requirements: You perform a threat analysis of your site, then you design a system that involves equipment and procedures, and finally you test it for adaptation. The system itself typically has a number of

Figure 5.1 Edinburgh Castle in Edinburgh, Scotland. Defending an asset with a multiple posture can reduce the likelihood of a successful attack; if one layer of defense fails, another layer of defense will hopefully prevent the attack, and so on.

elements that fall into the same essence that is at the heart of protection: *deter–detect–delay–respond.*

Security systems are often designed utilizing multiple barriers—"rings of protection"—encircling the protected asset. Layered barrier designs are advantageous when they require increased knowledge, skill, and talent to circumvent them. A group of attackers with the necessary skills must be assembled and since group secrecy is hard to maintain, the likelihood of being discovered is increased. Layered barriers also afford a greater time delay because each safeguard layer requires time to be circumvented. This helps to provide the necessary delay if the response time is relatively slow.

A system with protection in depth is one that has a number of protective devices in sequence. This is a great advantage because an adversary not only has to avoid or defeat one system, but also has to avoid or defeat several. "The actions and times required to penetrate each of these layers may not necessarily be equal, and the effectiveness of each may be quite different, but each will require a separate and distinct act by the adversary moving along the path."[3]

Each of these systems will require the adversary to use a separate and distinct act as he moves through. The effect produced on the adversary by a system that provides protection in depth over one that is very secure at one level is that it will increase uncertainty about the system, it will require more extensive preparations prior to attacking the system, and it will create additional steps where the adversary may fail or abort the mission.

The following critical building components should be located away from main entrances, vehicle circulation, parking, and maintenance areas. If this is not possible, harden as appropriate:

- Emergency generator, including fuel systems, day tank, fire sprinkler, and water supply
- Fuel storage
- Telephone distribution and main switchgear
- Fire pumps
- Building control centers
- Uninterrupted power supply (UPS) systems controlling critical functions
- HVAC systems if critical to building operation
- Elevator machinery and controls
- Shafts for stairs, elevators, and utilities
- Critical distribution feeders for emergency power

Protection Plans

Effective building security requires careful planning and design of the physical protection system—integrating people, procedures, and equipment, the foundation of all facility security operations. Protecting a building, its occupants, and related assets can pose a complex problem, and there is no perfect defense to all of the potential threats a target may face. Optimizing building security with respect to performance, cost, and efficiency ultimately requires compromise and balance in the application and consideration of people, procedures, and equipment. The fundamental aspects of building operations, however, are built upon three basic components: people, procedure, and technology. The combination of these elements contributes to overall organizational security as well as providing the basis for effective emergency preparation and response.[4]

As a security professional you know that people are the most important consideration for any security plan. There is no way people can be regarded as expendable assets. Yes, people can be replaced with new personnel, but they are an asset that has contributed to the overall success of the company and must be protected to the highest extent possible. A successful plan takes into consideration which departments need to be functional in the shortest amount of time possible and to quickly establish procedures to bring those operations back to full working order as soon as possible. This will help to ensure that procedures outlined in the plan take into account where the highest concentrations of priority operations need to be addressed.

The other category of people is those who provide protection. In the "people" element of security operations, the integration of people as a layer of security in the form of security guards and other security personnel is another consideration. The architectural layout of the building can be designed to influence the movement of people for rapid evacuation, limited congregation, increased visibility, and to limit the need for an abundant number of security personnel.

Security is a dynamic process, and in order for it to be effective it must be procedural in nature. For example, emergency response and business continuity plans must be developed well in advance of a critical incident and must define the plan of action or steps to be taken in a logical, orderly, and procedural manner. The development of procedures involves planning for such events as evacuation, emergency response, and disaster recovery in response to fires, natural disasters, and criminal intrusions.

Policies and procedures are then developed to assist with the proper response and recovery actions in the event that a crisis strikes. It should be noted that policies and procedures are guides to actions that should be taken in the event of an emergency and should not be strictly construed. No critical incident is the same, and it would be impossible to develop a procedure for every possible event that could occur. Procedures should be written so that they can be adaptable to any application.

The third element of the security operation, technology, involves hardware, electronics, and other equipment used in the security mission. In building security design, technologies can be built in or retrofitted into the existing structure to perform a variety of functions such as access control, surveillance, personnel screening, intrusion detection, and fortification. Technology has also advanced competitive intelligence and espionage. Global positioning systems and high-resolution surveillance are pushing security to new levels. There is technology now that, if an item is taken from a facility, that item can be tracked with the person that took it via GPS with a tiny sensor attached or scanned onto it.

These elements of security, people, procedures, and technology are interdependent because they rely on each other to be effective. For example, the behaviors and needs of people dictate what procedures and equipment may be deployed; procedures depend on people to be effective and equipment requirements depend on the particular procedures to be followed in a critical incident. A cost-effective and comprehensive plan necessitates a balance of these three elements, taking into account their particular contribution to the mission; one application may be security personnel intensive, where another may be equipment intensive.

Evacuation Drills

From the days when we were all in grade school we are all aware of the need to have fire drills. Every organization should have an emergency management plan developed in partnership with public safety agencies, including law enforcement and fire and local emergency preparedness agencies. The plan should address fire and natural and man-made disasters. An organization's plan should be tailored to address the unique circumstances and needs of the operation. For example, organizations on the east coast of the United States do not need to prepare for earthquakes, but all operations in California will put this as one of their top priority drills.

Staff training, particularly for those with specific responsibilities during an event, will include the combination of security personnel, facilities, and selected employees designated as floor fire wardens. Holding regularly scheduled practice drills, similar to the common fire drill, allows for plan testing, as well as employee and key staff rehearsal of the plan, and increases the likelihood for success in an actual event.

If the organization has a visitor management software program incorporated with access control, this will provide a system for **knowing who is in your building, including customers and visitors.**

In the United States, the FEMA Emergency Management Guide for Business and Industry outlines specific areas that need to be addressed and implemented during an evacuation of a facility:

1. **Decide in advance who has the authority to order an evacuation.** Create a chain of command so that others are authorized to act in case your designated person is not available. If local officials tell you to evacuate, do so immediately.
2. Identify **who will shut down critical operations and lock the doors**, if possible, during an evacuation.
    a. Choose employees most able to make decisions that emphasize personal safety first.

b. Train others who can serve as a backup if the designated person is unavailable.
   c. Write down, distribute, and practice evacuation procedures.
3. Locate and make copies of **building and site maps** with critical utility and emergency routes clearly marked.
   a. **Identify and clearly mark entry/exit points** both on the maps and throughout the building.
   b. **Post maps** for quick reference by employees.
   c. Keep copies of building and site maps with your crisis management plan and other important documents in your emergency supply kit and also at an off-site location.
   d. Make copies available to first responders or other emergency personnel.
4. **Plan two ways out** of the building from different locations throughout your facility.
5. Consider the feasibility of installing **emergency lighting** or plan to use flashlights in case the power goes out.
6. Establish a **warning system**.
   a. **Test systems** frequently.
   b. Plan to communicate with **people who are hearing impaired or have other disabilities** and those who do **not speak the local language**.
7. Designate an **assembly site**.
   a. Pick one location near your facility and another in the general area in case you have to move farther away.
   b. Talk to your people in advance about the importance of letting someone know if you cannot get to the assembly site or if you must leave it.
   c. Be sure that the assembly site is away from traffic lanes and is safe for pedestrians.
8. Try to **account** for **all workers, visitors, and customers** as people arrive at the assembly site.
   a. Take a head count.
   b. Use a prepared roster or checklist.
   c. Ask everyone to let others know if they are leaving the assembly site.
9. Determine **who is responsible for providing an all-clear** or return-to-work notification. Plan to cooperate with local authorities responding in an emergency.

10. Plan for **people with disabilities** who may need help getting out in an emergency.
11. If your business operates out of more than one location or has more than one place where people work, establish **evacuation procedures for each individual building**.
12. If your company is in a high-rise building, an industrial park, or even a small strip mall, it is important to **coordinate and practice with other tenants or businesses** to avoid confusion and potential gridlock.
13. If you **rent, lease, or share space with other businesses** make sure the **building owner** and other companies are committed to coordinating and practicing evacuation procedures together.

There are also special requirements if you are in a high-rise building, which is a building with more than seven floors:

1. Note where the closest **emergency exit** is.
2. Be sure you know **another way out** in case your first choice is blocked.
3. **Take cover** against a desk or table if things are falling.
4. **Move away** from file cabinets, bookshelves, or other things that might fall.
5. **Face away** from windows and glass.
6. **Move away** from exterior walls.
7. **Listen** for and follow **instructions**.
8. Take your **emergency supply kit**, unless there is reason to believe that it has been contaminated.
9. **Do not use elevators**.
10. Stay to the side while going down stairwells to allow emergency workers to come up.[5]

There may also be requirements to "shelter in place." There may be situations when it is best to stay where you are to avoid any uncertainty outside. There are other circumstances, such as during a tornado or a chemical incident, when specifically **how** and **where** you take shelter is **a matter of survival**. You should understand the different threats and **plan for all possibilities**.

FEMA has developed a system to put into place if you are instructed by local authorities to take shelter.[6]

Determine where you will take shelter in case of a **tornado warning:**

1. Storm cellars or basements provide the best protection.
2. If underground shelter is not available, go into an interior room or hallway on the lowest floor possible.
3. In a high-rise building, go to a small interior room or hallway on the lowest floor possible.
4. Stay away from windows, doors, and outside walls. Go to the center of the room. Stay away from corners because they attract debris.
5. Stay in the shelter location until the danger has passed.

If local authorities believe the air is badly contaminated with a chemical, you may be instructed to take shelter and "seal the room" (Figure 5.2).

The process used to seal the room is considered a temporary protective measure to create a barrier between your people and potentially contaminated air outside. It is a type of sheltering that requires preplanning:

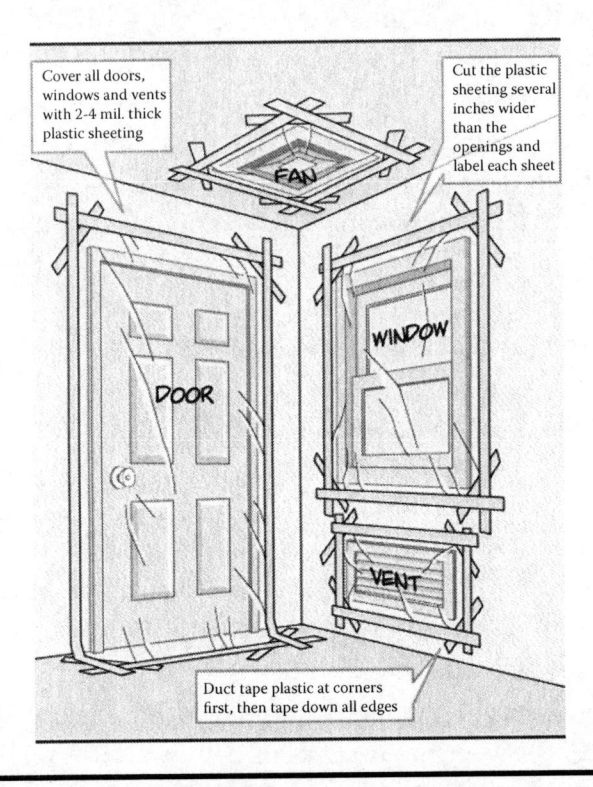

Figure 5.2 Shelter in place is a process used to seal a room as a temporary protective measure to create a barrier between you and potentially contaminated air outside. (Accessed May 2, 2012, from http://www.ready.gov/shelter)

1. **Identify a location to "seal the room" in advance.**
   a. If feasible, **choose an interior room**, such as a break room or conference room, with as **few windows and doors** as possible.
   b. If your business is located on more than one floor or in more than one building, **identify multiple shelter locations**.
2. **To "seal the room" effectively:**
   a. **Close** the business and bring **everyone inside**.
   b. **Lock** doors and **close** windows, air vents, and fireplace dampers.
   c. **Turn off** fans, air conditioning, and forced air heating systems.
   d. **Take your emergency supply kit** unless you have reason to believe that it has been contaminated.
   e. **Go into an interior room**, such as a break room or conference room, with few windows, if possible.
   f. **Seal** all windows, doors, and air vents with plastic sheeting and duct tape. Measure and cut the sheeting in advance to save time.
   g. Be prepared to **improvise** and use what you have on hand to **seal gaps** so that you create a barrier between yourself and any contamination.
   h. Local authorities may not immediately be able to provide information on what is happening and what you should do. However, you should **watch TV, listen to the radio, or check the Internet often for official news** and instructions as they become available.

Incident Response

An incident response plan takes its place beside business continuity and disaster-recovery plans as a key corporate document that helps guarantee that companies will survive whatever glitch, emergency, or calamity comes their way. According to George McBride, director of IT Risk Consulting,

> The typical response to trouble—the deer-caught-in-the-headlights look—is exactly why companies need such a plan. And while a business continuity plan aims to preserve operations in the face of adversity and a disaster recovery plan details what to do in case of a disaster, an incident response plan is broader, laying out how to respond to scenarios as diverse as data security breaches and network crashes.[7]

Plans and procedures, including recovery plans, emergency response, and evacuation, are deployed in response to different kinds of security and safety threats like earthquakes, explosions, lightning damage, and fires. Once the plans and procedures have been established, policies can be deployed to show how security personnel respond to these threats and to assist in recovery from other incidents that may occur.

Technology involves where and how screening of personnel and materials will be accomplished, and what kinds of systems and equipment will be used. The technology must be suitable for the operations or mission of the facility.

Some communities rely on businesses to generate jobs and tax revenue and to nurture an environment that is healthy and suitable. When a business protects itself from natural disasters, it also protects one of the community's most valuable assets. There is no way to prevent a natural disaster from occurring; however, you can take action to avoid the most devastating damage your business may face. The lack of knowledge and fear of the unknown about what to do in the case of emergency have caused businesses to fail or increase their cost due to an extended recovery time. Companies have to learn to identify the risk and hazards facing their business, to determine what types of training are useful, and to deal with and categorize the risk rationally. The goal of emergency preparedness is to be reasonably prepared and not to be swept up in a sea of confusion.

The best time to respond to a disaster is before it happens. A relatively small investment of time and money now may prevent severe damage and disruption of life and business in the future. Every area of the world is subject to some kind of disaster; you do not have to live on the coast to experience a severe event. Floods, hurricanes, earthquakes, ice storms, and landslides could happen anytime. With global warming and other unnatural events happening all over the world, we businesses must prepare for all possible occurrences. Everyone responds to disasters differently; however, business owners can take advantage of typical human behavior in the aftermath of an event. Seek out those individuals who rise above the chaos and get them trained and involved in recovery activities. This will improve the organization's chances for a shorter recovery time and a successful recovery effort.

The security professional needs to put together a plan of action and, in doing so, will identify areas that need to be addressed when an incident response is required. There are five topics that can be initiated and prepared for before an actual incident response is called upon:

1. **Identify what can happen.** If you have completed a security survey or risk assessment, then you have made determinations on what hazards are possible. However, although you cannot possibly anticipate what will happen in a crisis or during the aftermath, that does not mean that you cannot plan for one.
2. **Put together the team.** You need to put into place the team that will be utilized during an incident response and what roles each person will play when something happens.
3. **Have a communication plan.** When the decision is made to initiate the incident response, how will you get in touch with the team? The plan should include the individual contact information for team members that goes well beyond office e-mail addresses and phone extensions. There should also be home phone numbers and e-mails along with mobile phone numbers.
4. **Identify who does what and when.** Good incident response plans do not just name the members of the response team; rather, they lay out who will have which responsibilities and authority so that they can get right to work.
5. **Test your plan.** You do not want to find holes and glitches in your incident response plan when you are dealing with a fire, burst water pipe, or a bomb threat. That is why it is so important to test it ahead of time.

Penetration Tests

From an IT standpoint, the idea of a penetration test is immediately connected to testing the network defenses by attempting to access a system from the outside using hacker techniques. However, from the physical security aspect, a much more "in your face" and personal approach needs to be addressed. There are several variations of physical penetration methods used by perpetrators, including dumpster diving, lock picking, social engineering, physical access compromise, and "simulated sabotage." While these techniques may seem extreme, it is important to remember that bad guys do not follow the rules and they do not play nice.

"Tests should be designed with the purpose of the system and the specific type of prevention desired in mind. In addition, threats evolve so quickly that penetration testing should be a regular occurrence, not just a one-time event."[8]

There are many benefits to conducting a penetration test on your facility. It will identify vulnerable areas that need to be addressed immediately. In order to determine the effectiveness of the security apparatuses in place, a penetration test is essential. It is easy to say that we have outstanding security personnel and a security program in place but it is another thing to verify and confirm this.

- Will the guard at the front desk actually look for a perpetrator or an individual without a badge?
- Can the guard spot a forged badge or whether the picture on the badge is that of the individual?
- Will employees stop and question someone who looks out of place inside the building? Will they contact security that there is someone on their floor who does not belong?
- Will employees hold open a door for someone trying to enter through an employee-only entrance or will they make the person use his or her access card? Does the employee-only entrance require a dual-technology badge and pin number?

Having an outside entity attempt the penetration is the best method. They look for easy accesses into the facility or they try to mingle in with the crowd during morning hours when everyone is coming into work. If your facility utilizes a contract guard force, this can be done with the coordination of their upper management. There are also several companies that specialize in penetration testing.

In a penetration test that I was associated with, the penetration tester was given several fake badges that had a picture on the badge that was similar to his looks but was obviously not him. If the guard was to scrutinize the badge and compare it with the holder, it would be obvious. The penetration tester was instructed to show the badge to the receptionist and attempt to enter as if he belonged. The first attempts were successful and the penetration tester gained entry. After the receptionists were made aware and counseled by management on their assigned duties and responsibilities, a second test was conducted 2 months later and this time a 70% reduction in successful penetrations resulted. Once the receptionists and guards were made aware of tests and that their response would affect their performance rating, they became true guardians of the gate.

This test was done to see if the facility could stop a person from gaining entry from a physically manned entry/exit point. But from the perspective of an all-inclusive test, there needs to be more emphasis on an all-out penetration attack.

Another perimeter area to look at concerns whether, when construction is going on, a perpetrator can put on a hardhat and walk in with the construction crew, then discard the construction look and walk into the facility like he or she belongs. How are construction, contractors, and vendors controlled?

Are secured areas truly secured? Are there layers of security within the structure or is it considered an M&M's® candy—hard on the outside but soft on the inside? These facilities place all the emphasis on the outer security perimeter, but once you have navigated through into the building, there are virtually no security measures.

A penetration test can determine if security measures are enforced within the building. Are all employees required to wear identification badges in plain sight while they are inside the building? Are there security awareness posters displayed?

After an intruder has entered through the perimeter, can he walk into areas within the facility and go through an employee area without being disturbed? Will anyone confront and question his reason for being in the area?

The penetration test will identify vulnerabilities at the perimeter, but what about exiting the facility? Are there any safeguards set up to deter the removal of classified information or property from the facility?

> Does the company check for theft of sensitive material when employees exit the facility? Are laptop computers or other portable devices registered and checked when entering and exiting the building? Are security guards trained not only on what types of equipment and information to look for, but also on how equipment can be hidden or masked and why this procedure is important?[9]

Other methods used in penetration testing include a social engineering scheme, where the attacker relies on human nature to gain access to unauthorized network resources. This could be in the form of eavesdropping or "shoulder surfing" (looking over your shoulder) to obtain access information. The basic goals of social engineering are the same as hacking in general: to gain unauthorized access to systems or information in order to commit fraud, network intrusion, industrial espionage, or identity theft, or simply to disrupt the system or network. The Computer Security Institute[10] gives an example of how to use social engineering to gain critical information:

The facilitator of a live Computer Security Institute demonstration neatly illustrated the vulnerability of help desks when he "dialed up a phone company, got transferred around, and reached the Help Desk":

Facilitator: Who's the supervisor on duty tonight?
Employee: Oh, it's Betty.
Facilitator: Let me talk to Betty. [He is transferred.] Hi, Betty, having a bad day?
Betty: No, why?
Facilitator: Your systems are down.
Betty: My systems aren't down; we're running fine.
Facilitator: You'd better sign off.

   [Betty signs off.]

Facilitator: Now sign on again.

   [Betty signs on again.]

Facilitator: We didn't even show a blip; we show no change. Sign off again.

   [Betty signs off again.]

Facilitator: Betty, I'm going to have to sign on as you here to figure out what's happening with your ID. Let me have your user ID and password.

So, this senior supervisor at the Help Desk told the caller her user ID and password.

The natural human willingness to accept someone at his or her word leaves many of us vulnerable to attack. There is an old saying, "In God we trust; all others we verify." This needs to be instilled in staff. All employees should be trained on how to keep confidential data safe. The simple statement, "I am sorry, but I do not know who you are and I will not be providing that information over the phone," will provide for a level of security.

Another of the penetration tests is the simple task of checking your trash. "Dumpster diving" is the practice of searching through the trash of an individual or business in attempts to obtain something useful (Figure 5.3). It can also include data aggregation by looking for passwords written on sticky notes, unwanted files, letters, memos, photographs, IDs, and other paperwork that has been found in dumpsters. This oversight is a result of many people not realizing that sensitive items like passwords, credit card numbers, and personal information they throw in the trash could be recovered anywhere from the dumpster to the landfill.

Every business has information that is confidential and must be disposed of properly. Carelessly discarded correspondence, financial statements, medical records, credit card statements, photocopies, and computer printouts are all easily removed from the trash. If these data get into the wrong hands, this can cause acute embarrassment, financial loss, or legal action. Loss of government data can result in a serious breach of national security. Your personal information in the wrong hands can be just as damaging.

Figure 5.3 You will be surprised, alarmingly so, as to how much proprietary information can be acquired by undesirables simply going through your trash.

Protecting your privacy is a vital necessity and shredding has become standard practice in offices as well as homes.

There are several methods for proper destruction of information. Your company can contract with a licensed and bonded shredding company that will come to your site with a mobile shredding truck and dispose of your sensitive information while you watch and verify the destruction (photo) or you can shred on site depending on the volume of information that needs to be destroyed. Shredding services can also have the capacity to destroy hard drives and physical components.

Table 5.1 lists some common intrusion tactics and strategies for prevention.[11]

Access Control Violation Monitoring

When doors are not physically controlled by a guard, there is a tendency for personnel to violate entry procedures. Violation of access control systems, when controlled by card reader, may occur by "tailgating" or "piggybacking," where an authorized employee with a valid entry card is accompanied by a closely spaced nonauthorized perpetrator or an authorized employee who inadvertently failed to follow proper entry procedures without considering the security consequences.

If your operation has an employee entrance that does not have a mantrap or turnstile entry system and only has a single door, there are products available to announce when a tailgate has occurred. This could be a buzzer local to the door that sounds to alert a valid cardholder to challenge someone tailgating behind him or her.

In higher end security systems, the alarm may be used to alert a control room operator and trigger live closed circuit television (CCTV) images, allowing immediate action to be taken. When coupled with a modern integrated security management system, a full alarm event history can be produced indicating the date, time, and location of the alarm; the cardholder who allowed someone to tailgate; and the digital CCTV images of the person tailgating.

This is why a protection-in-depth approach is necessary. If a perimeter door is compromised and an individual has gained entry into the facility, the area that has been entered will not be a high-value area and a response team can cordon off the area that has been breached and make contact with the violator.

Maintaining an audit trail of improper entry attempts or entry violations (allowing tailgating) is a way to identify employees who need additional

Table 5.1 Common Intrusion Tactics and Strategies

Area of Risk	Hacker Tactic	Combat Strategy
Phone (Help Desk)	Impersonation and persuasion	Train employees/Help Desk never to give out passwords or other confidential information by phone
Building entrance	Unauthorized physical access	Tight badge security, employee training, and security officers present
Office	Shoulder surfing	Do not type in passwords with anyone else present (or if you must, do it quickly!)
Phone (Help Desk)	Impersonation on Help Desk calls	All employees should be assigned a PIN specific to Help Desk support
Office	Wandering through halls looking for open offices	Require all guests to be escorted
Mail room	Insertion of forged memos	Lock and monitor mail room
Machine room/phone closet	Attempting to gain access, remove equipment, and/or attach a protocol analyzer to grab confidential data	Keep phone closets, server rooms, etc., locked at all times and keep updated inventory on equipment
Phone and PBX	Stealing phone toll access	Control overseas and long-distance calls, trace calls, refuse transfers
Dumpsters	Dumpster diving	Keep all trash in secured, monitored areas; shred important data; erase magnetic media
Intranet/Internet	Creation and insertion of mock software on Intranet or Internet to "snarf" passwords	Continual awareness of system and network changes; training on password use
Office	Stealing sensitive documents	Mark documents as confidential and require those documents to be locked up

(continued)

Table 5.1 Common Intrusion Tactics and Strategies (continued)

Area of Risk	Hacker Tactic	Combat Strategy
General—psychological	Impersonation and persuasion	Keep employees on their toes through continued awareness and training programs

training on proper security entry requirements or—more drastically for continual violations—having their supervisor notified and, if all else fails, having their badge revoked and requiring them to be escorted all day. This is a drastic move but they will only need to be escorted once by a fellow worker before they get the message and will adhere to proper security procedures.

References

1. Cramsession.com. 2007. Building a defense in depth toolkit. Retrieved March 1, 2007. http://www.cramsession.com/articles/get-article.asp?aid=1105
2. Garcia, M. L. 2006. *Vulnerability assessment of physical protection systems, 35*. Burlington, MA: Elsevier Butterworth-Heinemann.
3. Ibid., 41.
4. Ibid., 9–14.
5. http://www.ready.gov/america/makeaplan/highrise.html
6. http://www.fema.gov/plan/prevent/saferoom/index.shtm
7. Pratt, M. 2007. Five tips for building an incident response plan. Computerworld. http://www.computerworld.com/action/article.do?command=viewArticleBasic&articleId=9019558
8. Tiller, J. 2004. *The ethical hack: A framework for business value penetration testing*. Boca Raton, FL: Auerbach Publications/CRC Press.
9. Tipton, H., and M. Krause. 2006. *Information security management handbook*, 181. Boca Raton, FL: CRC/Taylor & Francis Group.
10. Granger, S. 2001. Social engineering fundamentals, part I: Hacker tactics. http://www.securityfocus.com/infocus/1527
11. Granger, S. 2002. Social engineering fundamentals, part II: Combat strategies. http://www.securityfocus.com/infocus/1533

Chapter 6

Perimeter Protection

Paul R. Baker

Crime Prevention through Environmental Design (CPTED)

CPTED is a crime reduction technique that has several key elements applicable to the analysis of the building function and site design against physical attack. It is used by architects, city planners, landscapers, interior designers, and security professionals with the objective of creating a climate of safety in a community by designing a physical environment that positively influences human behavior.

CPTED concepts have been successfully applied in a wide variety of applications, including streets, parks, museums, government buildings, houses, and commercial complexes. The CPTED process provides direction to solve the challenges of crime with organizational (people), mechanical (technology and hardware), and natural design (architecture and circulation flow) methods.

The CPTED approach recognizes the building environment's design or redesign use. The emphasis of security design falls on the design and use of space, a practice that is different from traditional hardening. Traditional target hardening focuses predominantly on denying access to a crime target through physical or artificial barrier techniques such as locks, alarms, fences, and gates. The traditional approach tends to overlook opportunities for natural access control and surveillance.

The environmental security design is based on three functions:

- Designation: What is the purpose or intended use of the space?
- Definition: How is the space defined? Is it work space, a meeting area, warehouse space, etc.?
- Design: Is the space defined to support the prescribed or intended behaviors?

CPTED can be integrated into expansion or reconstruction plans for existing buildings as well as plans for new buildings. Applying CPTED concepts from the beginning usually has minimal impact on costs and the result is a safer facility (Figure 6.1).

Landscape design features should be used to create the desired level of protection without turning the facility into a fortress. Elements such as landforms, water features, and vegetation are among the building blocks of attractive and welcoming spaces, and they can also be powerful tools for enhancing security. During site planning it would be beneficial to consider and install these techniques from a cost-savings approach. The earth movers, graders, and landscapers have all been budgeted, so why not use CPTED techniques to supplement security concerns?

Figure 6.1 This building has an open entranceway with lots of use of glass for clear lines of sight and natural surveillance of the entrance. Minimal foliage offers no hiding places for individuals. Parking is not right up against the building for an added layer of security.

Stands of trees, natural earthen berms, and similar countermeasures generally cannot replace setbacks, but they can offer supplementary protection. With careful selection, placement, and maintenance, landscape elements can provide visual screening that protects employee gathering areas and other activities from surveillance without creating concealment for covert activity. However, dense vegetation in close proximity to a building can screen illicit activity and should be avoided. Additionally, thick ground cover or vegetation over 4 inches tall can be a security disadvantage; in setback clear zones, vegetation should be selected and maintained with eliminating concealment opportunities in mind. Similarly, measures to screen visually detractive components such as transformers, trash compactors, and condensing units should be designed to minimize concealment opportunities for people and weapons.

The following is from The Seven Qualities for Well-Designed, Safer Places:[1]

> Avoid using elements that create a poor image or a fortress-like appearance. Integrate any necessary security features into buildings or public spaces by designing them to be intrinsic, unobtrusive or a positive visual feature. Possible design techniques include:
>
> - Treating gates and grilles as public art
> - Making perimeter fences look attractive by allowing visibility through the fences, including simple design motifs or combining them with a hedge of thorny shrub varieties that can "target harden" boundary treatment
> - Using open grilled designs or internal shutters instead of roller-shutter blinds
> - Using different grades of toughened or laminated glass as a design alternative to various types of grilles

Many CPTED crime prevention techniques are commonsense approaches. For example, businesses are encouraged to direct all visitors through one entrance that offers contact with a receptionist. This person can determine the purpose of the visit and the destination and provide sign in/sign out and an ID badge prior to granting building access. These measures are nothing new to the retail business world. This approach encourages employees to make personal contact with everyone entering the store in order to keep track of people who want to be invisible and do not want to attract the attention of store employees while they perpetrate crimes.

Other CTPED concepts include the idea that a standard front with windows overlooking sidewalks and parking lots is more effective than encircling the facility with cyclone fences and barbed wire. A communal area with picnic seating, in which activities happen frequently, has a greater deterrent effect. Trees also help, as they make shared areas feel safer. Access matters too; defensible spaces should have single egress points so that potential intruders are afraid of being trapped. It has been found, for example, that CCTV cameras only deter crime in facilities such as parking lots where there is a single exit.[2]

Protecting with CPTED Concepts

The attacks on the World Trade Center, the Pentagon and the Oklahoma City bombings are forever etched as terrorism landmarks in our collective memory. Terrorism represents a real threat for our society and to our peace of mind. The face of terrorism is undergoing systemic changes as the level of terrorist sophistication increases with the availability of knowledge and materials with which to carry out these acts of violence. Knowledge about bombs and terror has proliferated to the point that virtually any terrorist or criminal can find the information needed to build virtually any kind of explosive device.

Timothy McVeigh, who blew up the Alfred R. Murrah Federal Building, stated in an interview shortly after his arrest that he picked that particular building because "it was more architecturally vulnerable." Who would have ever thought that a rental truck and a load of manure could be so deadly?

What can we do to diminish the threats to and losses of persons, information, and property? How do you reduce the opportunity and fear of terrorism in the built environment with CPTED?

In June 1995, President Bill Clinton mandated basic standards of security for all federal facilities. The mandate states that each federal building is to be upgraded to the minimum-security standards recommended for its audited security level by the Department of Justice. In November 2001, President George W. Bush signed a bill federalizing airport security screeners and antiterrorism legislation and empowering law enforcement in the military to take preventive actions.

Before the US Marshals Service conducted a vulnerability assessment in the wake of the Murrah Building bombing in 1995, there were no government-wide standards for security at federal buildings. The US Marshals

Service Building Security Study developed 52 standards, primarily covering perimeter security, entry security, interior security, and security technology planning. Each federal building was rated within five levels of security based on facility size, facility population, and level of public access: Level I was a minimum security building and level V a defense plant or nuclear facility. Most courthouses with a multitenant, multistory building are considered level III and require shatter-resistant glass, controlled parking, 24-hour CCTV monitoring and videotaping, x-ray weapon and packet screening, and photo identification systems.

The General Services Administration (GSA) security standards encourage a defensible space/crime prevention through environmental design approach to clearly define and screen the flow of persons in vehicles through layering from public to private spaces. Edges and boundaries of the property should clearly define the desired circulation patterns in movements. The screening and funneling of persons through screening techniques is an effort to screen the legitimate users of the building from illegitimate users who might look for opportunities to commit crime, workplace violence, or acts of terrorism.

The result of 1 year of work by the GSA panel is a set of criteria covering four levels of protection for every aspect of security in the US Marshals report, which made a large number of recommendations for both operational and equipment improvements. The GSA security standards address the functional requirements and desired application of security glazing, bomb-restraint design and construction, landscaping and planting design, site lighting, natural and mechanical surveillance opportunities good site line, no blind spots, window placement, and proper application of CCTV. These recommendations were further subdivided according to whether they should be implemented for various levels of security (e.g., a level I facility might not require an entry control system, while a level 4 facility would require electronic controls with CCTV assessments).

What follows are some of the general guidelines that the architect, engineers, and security team should address for major renovations or new construction on any federal building. Although not required, the exercise of due diligence suggests that state governments and commercial businesses also consider these standards in new construction as a comparable reference point or standard of care.

The GSA security standards address the following key areas that are applicable to most comparable buildings:

1. Perimeter and exterior security
    a. Parking area and parking controls
    b. CCTV monitoring
    c. Lighting to include emergency backup
    d. Physical barriers
2. Entry security
    a. Intrusion detection systems
    b. Upgrade to current life safety standards
    c. Screen mail person's packages
    d. Entry control with CCTV electric door strikes
    e. High-security locks
3. Interior security
    a. Employee ID visitor control
    b. Control access to utilities
    c. Provide emergency power to critical systems
    d. Evaluate location of day care centers
4. Security planning
    a. Evaluate the locations of tenant agencies and leased buildings and assess security needs and risks
    b. Install security film on exterior windows
    c. Review established blast standards for current relevant projects in new construction
    d. Consider blast-resistant design and street setbacks for new construction of high-risk buildings (level III or IV)

The GSA criteria take a balanced approach to security, considering cost effectiveness, acknowledging acceptance of some risk, and recognizing that federal buildings should not be bunkers or fortress-like, but rather open, accessible, attractive, and representative of the democratic spirit of the country. The guidelines suggest prudent rather than excessive security measures are appropriate in facilities owned by and serving the public.

CPTED process provides a holistic mythology to meet the challenges of crime in terrorism with three approaches: organizational methods (people: security staff, capable guardians), mechanical methods (technology: hardware, barriers, hardening), and natural design methods (architectural, design, and circulation movement flow). Property owners must determine the assets they want to protect, the level of risk they are willing to assume, and how much they can afford to spend protecting their sites, facilities, employees, and other occupants.[3] A comprehensive building security plan integrates three elements: design, technology, and operations—each consisting of policies and procedures. These elements are most effective when used together and are appropriate to protect structures against terrorism, natural disaster, crime, and workplace violence.

For example, if one of the vulnerabilities of a threat analysis for a government building is the challenge of a truck bomb and the goal is to distance a potential bomb from the facade of the building, then the CPTED approach would propose careful considerations of the following:

- Where is the parking place?
- How does service delivery get screened and controlled?
- How do pedestrians flow into the building?
- How many entrances are there for the public, staff, and service?
- Is there one main entrance for the public?
- How much distance is the exterior path of travel from the street, pedestrian plaza, to the building facade?
- Do all four facades have setbacks from the street?
- What is the most appropriate bollard system or vehicle barrier system?
- Do bollards or planters create blind spots or sleeping places for homeless persons and street criminals?
- Does a threat exist from bicycle and motorcycle bombers, thus requiring a smaller barrier net?
- Does surveillance from the building to the street remain unobstructed?
- Do landscaping and plantings obstruct views?
- Do barriers hinder accessibility by persons with disabilities?
- Where do private or public security forces patrol?
- Is the structure of the building designed with structural redundancy?
- Does the building become a less appealing target by layers of buffer zones that make it more difficult for an intruder to reach the intended target?
- If the building is at high risk of a bomb threat, have the structural components been designed to allow for the negative pressure effects of an explosion?
- Does lighting around the property provide a uniform level of light to resist shadows and hiding places?
- Are there CCTVs in places of extraordinary activity to detect inappropriate behavior and record and monitor that activity?
- Does the building have a consistent and comprehensive weapon-screening program for the building users, staff, packages, and mail?

- Does the property use security layering to create a sense of boundary of the property, the building, and specific points within the building?
- Do management and maintenance practices and policies support security operations, the use of security staff, monitoring devices, weapon screening procedures for people and property, the screening of employees' backgrounds, and the physical upkeep of the premise?

It is evident that a lot of thought and money goes into making a building secure. However, neither an architect nor a security director can change human nature, and criminal acts will be perpetrated in spite of the best laid plans. Our building environment cannot be defended against every potential threat. No building security system could have prevented the acts of terrorism of 9/11 or the bombings of our embassies or courthouses. But there are many active steps that can be taken to reduce the opportunities for and fear of crime and to increase our awareness of the threats.

Many CPTED crime prevention techniques are commonsense approaches. For example, businesses are encouraged to direct all visitors through one entrance that offers contact with a receptionist who can determine the purpose of the visit and the destination, and provide sign in/sign out and an ID badge prior to building access. These measures are nothing new to the retail business world. This approach encourages employees to make personal contact with everyone entering the store in order to keep track of people who want to be invisible and do not want to attract the attention of store employees while they perpetrate crimes.

Other measures to enhance the security envelope on the perimeter are barriers, fences, gates, walls, and lighting.

Barriers

Barriers can comprise natural or man-made elements. The idea is to define an area that is designated to impede or deny access. A natural barrier can be a river, dense growth, a culvert, or a ditch. Man-made or structural barriers can be a wall, a fence, doors, or the building itself. Walls, fences, and gates have long been designated as the first line of defense for a facility. There are a multitude of barriers and they provide the same objective: keep intruders out, delay them, and keep them at a safe distance. However, with sufficient time and effort, any barrier can be breached. Thus, the principal objective is to delay the intruder until law enforcement or a security team can respond.

Fences

As shown in Figure 6.2(a)–(d), fences are a perimeter identifier designed and installed to keep intruders out. However, most organizations do not like the feeling of a fenced-in compound and look for other remedies to secure their property. Depending on the organization, location, and funding, a fence can consist of many variations and levels of protection.

The most commonly used fence is the chain-link fence and it is the most affordable. The standard is a 6-foot high fence with 2-inch mesh square openings. The material should consist of 9-gauge vinyl or galvanized metal. You may encounter 9 gauge being installed at your own home as a typical fence material. Additionally, it is recommended to place barbed wire strands angled out from the top of the fence at a 45° angle and away from the protected area with three strands running across the top. This will provide for a 7-foot fence. There are several variations of the use of "top guards" using V-shaped barbed wire or the use of concertina wire as an enhancement, which has been a replacement for more traditional three-strand barbed wire "top guards."

(a)

Figure 6.2(a) (a) Fencing with barbed wire and razor wire; (b) chain-link fencing at a school; (c) decorative wrought iron fencing; (d) barbed wire mounted on a traditional concrete wall. Modern high-tech fence detection sensors can be used in conjunction with traditional chain-link fencing for added perimeter security. **(continued)**

(b)

Figure 6.2(b) (continued) (a) Fencing with barbed wire and razor wire; (b) chain-link fencing at a school; (c) decorative wrought iron fencing; (d) barbed wire mounted on a traditional concrete wall. Modern high-tech fence detection sensors can be used in conjunction with traditional chain-link fencing for added perimeter security. (continued)

The fence should be fastened to ridged metal posts set in concrete every 6 feet with additional bracing at the corners and gate openings. The bottom of the fence should be stabilized against intruders crawling under by attaching posts along the bottom so as to keep the fence from being pushed or pulled up from the bottom. If the soil is sandy, the bottom edge of the fence should be installed below ground level.

For a maximum security design, the use of double fencing with rolls of concertina wire positioned between the two fences is the most effective deterrent and cost-efficient method. In this design an intruder now is required to use an extensive array of ladders and equipment to breach the fences.

Most fencing is largely a psychological deterrent and a boundary marker rather than a barrier, because in most cases such fences can be rather easily penetrated unless added security measures are taken to enhance the security of the fence. Sensors can be attached to the fence to provide electronic monitoring of cutting or scaling of the fence.

Gates

Gates exist to facilitate and control access. Gates need to be controlled to ensure that only authorized persons and vehicles pass through. A variety of controls are used. It is best to minimize the number of gates and access points because any opening is always a potential vulnerability.

(c)　　　　　　　　　　　　　　　　(d)

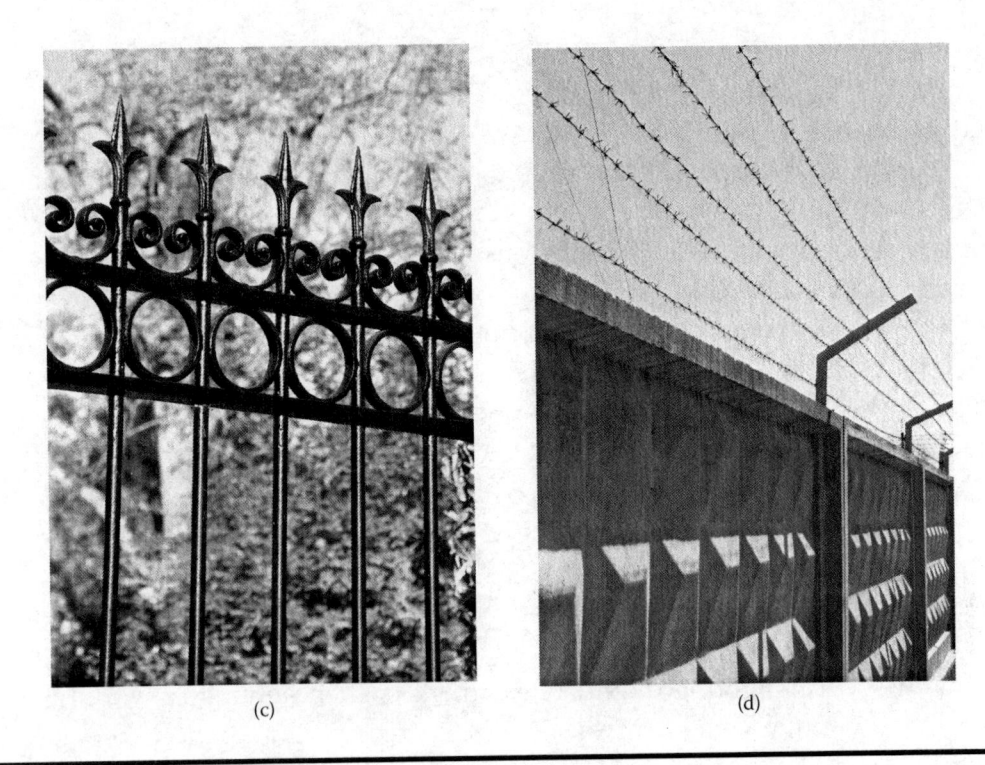

Figure 6.2(c, d) (continued) (a) Fencing with barbed wire and razor wire; (b) chain-link fencing at a school; (c) decorative wrought iron fencing; (d) barbed wire mounted on a traditional concrete wall. Modern high-tech fence detection sensors can be used in conjunction with traditional chain-link fencing for added perimeter security.

Each gate requires resources whether it uses electronic access control or a guard. The fewer the number of entry points, the better the control of the facility.

Walls

Walls serve the same purpose as fences. They are man-made barriers but generally are more expensive to install than fences. Common types of walls are block, masonry, brick, or stone. Walls tend to have a greater aesthetic value, appealing to those who prefer a more gentle and subtle look. Regardless of the type of wall used, its purpose as a barrier is the same as that of a fence. To be most effective, walls ought to be 7 feet high with three to four strands of barbed wire on top. This will help deter scaling. Walls also have a disadvantage in that they obstruct the view of an area. Chain-link and wire fencing allow for visual access from both sides.

Lighting

Security lighting can be provided for overall facility illumination along the perimeter to allow security personnel to maintain a visual assessment during times of darkness. It may provide both a real and psychological deterrent against intruders who will attempt to use the cover of darkness as a means of entry into a compound, parking lot, or facility. Lighting should enable security personnel and employees to notice individuals at night at a distance of 75 feet or more and to identify a human face at about 33 feet. These are distances that will allow the security personnel to avoid the individuals or take defensive action while still at a safe distance.

Security lighting increases the effectiveness of guard forces and closed circuit television by increasing the visual range of the guards or CCTV during periods of darkness. It also provides increased illumination of an area where natural light does not reach or is insufficient. Lighting also has value as a deterrent to individuals looking for an opportunity to commit crime. Normally, security lighting requires less intensity than lighting in working areas. An exception is at doorways, where increased illumination is required (Figure 6.3).

Lighting is relatively inexpensive to maintain and may reduce the need for security personnel while enhancing personal protection by reducing opportunities for concealment and surprise by potential attackers. Overall,

Figure 6.3 Proper lighting around doors and entrances is important for security as well as safety and to protect against liability. (Photo by Daniel J. Benny.)

sufficient lighting will be required at entry control points to ensure adequate identification of personnel. Also, wherever practical, place lighting devices as high as possible to give a broader and more natural light distribution. This requires fewer poles and is more aesthetically pleasing than standard lighting.

Types of Lighting Systems

The type of site lighting system used depends on the overall security requirements. Four types of lighting are used for security lighting systems:

- **Continuous lighting** is the most common security lighting system. It consists of a series of fixed lights arranged to flood a given area continuously during darkness with overlapping cones of light.
- **Standby lighting** has a layout similar to continuous lighting; however, the lights are not continuously lit, but rather are either automatically or manually turned on when suspicious activity is detected or suspected by the security personnel or alarm systems.
- **Movable lighting** consists of manually operated, movable searchlights that may be lit during hours of darkness or only as needed. The system normally is used to supplement continuous or standby lighting.
- **Emergency lighting** is a backup power system of lighting that may duplicate any or all of the preceding systems. Its use is limited to times of power failure or other emergencies that render the normal system inoperative. It depends on an alternative power source such as installed or portable generators or batteries. Consider emergency/backup power for security lighting as determined to be appropriate (Figure 6.4).

Depending on the nature of the facility, protective lighting will be deployed to illuminate the perimeter of the facility along with any outside approaches. It will also be utilized in order to concentrate on the inner area and the buildings within the perimeter. The Code of Federal Regulations[4] lists a specific requirement of 0.2 foot candles (fc) for lighting protected areas within a perimeter: "Isolation zones and all exterior areas within the protected area shall be provided with illumination sufficient for the monitoring and observation requirements…but not less than 0.2 fc measured horizontally at ground level." For side landscapes and roadways, 0.5 fc is acceptable.

Figure 6.4 Emergency power as backup for essential lighting, including exit signs, is an essential consideration. (Photo by Daniel J. Benny.)

But from the standpoint of a regular security profession, who can determine what exactly a foot candle is? The basic idea for perimeter lighting (0.5 fc) will equate to using a 40 W bulb in a 12 × 12 room. It gives off enough light to see, as it is a soft amber glow, but it will not totally illuminate the entire room.

Types of Lights

There are several types of light that can be used within the protected area. They include fluorescent, mercury vapor, sodium vapor, and quartz lamps.

- **Fluorescent** lights are highly efficient and cost effective. However, they are temperature sensitive and are not considered an effective outdoor lighting system. This light is better suited for inside buildings and facilities.
- **Mercury vapor** lights are the preferred security light, dispersing a strong white-bluish cast. They have an extended lamp life; however, the downside is they take an amount of time to light fully when activated—such as the lights at a stadium or a street light.
- **Sodium vapor** lights provide a soft yellow light and are more efficient than mercury vapor. This light is used in areas where fog could be a problem (Figure 6.5).
- **Quartz lamps** emit a very bright white light and come on immediately. They typically provide high wattage, from 1500 to 2000 W, and can be used on perimeters and troublesome areas where high visibility and a daylight scene are required.

Figure 6.5 Sodium lights help to cut through the fog—in this case, on a giant sea crane at a shipyard.

According to the American Institute of Architects,[5] interior lighting levels for elevators, lobbies, and stairwells range from 5 to 10 fc; exterior lighting requirements vary for different locations. Common lighting levels include the following:

Building entrances (5 fc)
Walkways (1.5 fc)
Parking garages (5 fc)
Site landscape (0.5 fc)
Areas immediately surrounding the building (1 fc)
Roadways (0.5 fc)

Adequate lighting for monitoring activities is important. In addition, lighting serves as a crime deterrent and discourages unwanted visitors while giving the building occupants a sense of security and safety. Lights used for CCTV monitoring generally require at least 1 to 2 fc of illumination, whereas the lighting needed for safety considerations in exterior areas such as parking lots or garages is substantially greater (at least 5 fc) (see Figure 6.6).

Figure 6.6 While the building itself and its immediate vicinity are brightly lit, the parking lot lighting is insufficient.

Infrared Illuminators

The human eye cannot see infrared light. Most monochrome (black/white) CCTV cameras can. Thus, invisible infrared (IR) light can be used to illuminate a scene, which allows night surveillance without the need for additional artificial lighting. IR beam shapes can be designed to optimize CCTV camera performance and can provide covert surveillance, with no visible lighting to alert or annoy neighbors. This is extremely effective in low-light areas and can provide the monitoring guard the ability to see in the dark. Alternatively, where light is not required but the ability to capture images might be, an infrared camera can serve the same purpose (Figure 6.7).

Figure 6.7 An infrared camera.

References

1. http://www.justice.govt.nz/publications/global-publications/n/national-guidelines-for-crime-prevention-through-environmental-design-in-new-zealand-part-1-seven-qualities-of-safer-places-part-2-implementation-guide-november-2005/the-seven-qualities-for-well-designed-safer-places
2. Jeffrey, R. 1971. CPTED.
3. Nadel, B. 2004. *Building security: Handbook for architectural planning and design*. New York: McGraw–Hill.
4. Code of Federal Regulations. http://books.google.com/books?id=tyQ5AAAAIAAJ&pg=PA577&lpg=PA577&dq=Isolation+zones+and+all+exterior+areas+within+the+protected+area&source=bl&ots=AFJ8NcCDjR&sig=mfvd-EhkiBJ281J0EGMHIUP5cU0&hl=en&sa=X&ei=XiTeT9bcDoKi8gSs2YXJCg&sqi=2&ved=0CFEQ6AEwAA#v=onepage&q=Isolation%20zones%20and%20all%20exterior%20areas%20within%20the%20protected%20area&f=false
5. The American Institute of Architects. 2004. *Security planning and design: A guide for architects and building design professionals*. Hoboken, NJ: Wiley Publishing.

Chapter 7

Access Control

Paul R. Baker

Access Control

The primary function of an access control system (ACS) is to ensure that only authorized personnel are permitted inside the controlled area. This can be accomplished by using a guard force, electronic card readers, biometrics, portals, mantraps, or anything that will control the access into your facility. This can also include the regulation and flow of materials into and out of specific areas. Persons subject to control can include employees, visitors, customers, vendors, and the public. Access control measures should be different for each application to fulfill specific security, cost, and operational objectives. The goal of entry control is

1. To allow only authorized personnel the ability to enter and exit the facility
2. To detect and prevent the entry or exit of unauthorized material
3. To provide information to security personnel in order to make an assessment and determine a response

Control can begin at the facility property line to include such areas as parking lots. Here a transmitter is installed on the vehicle windshield, similar to an "Easy Pass" that is used throughout the United States at tolls so as to speed up traffic flow. This device can be read by a sensor that will act as a card reader and permit or deny vehicle access to a parking or garage area.

Within the facility, any area can be controlled at the discretion of management. However, control is normally applied to be consistent with identified risk and the protective value that is desired. Protected areas can include street-level entrances, lobbies, loading docks, elevators, and sensitive internal areas containing assets such as customer data, proprietary information, and classified information.

The goal of an access control program is twofold: (1) to allow authorized personnel into areas within the facility that management has designated that they have the need to enter, and (2) to limit the opportunity for a crime to be committed. If the potential perpetrator of a crime cannot gain access to financial assets, data files, computer equipment, programs, documentation, forms, operating procedures, and other sensitive material, the ability to commit a crime against the institution is minimized. Thus, only identified, authorized personnel should be permitted access to restricted areas.

The basic components of an automated ACS include card readers, electric locks, alarms, and computer systems to monitor and control the ACS (Figure 7.1).

In order for the system to identify an authorized employee, an ACS needs to have some form of enrollment station that is used to assign and activate an access control device. Most often a badge is produced and issued with the employee's identifiers and the enrollment station gives the employee specific areas that will be accessible. Typically, a new employee

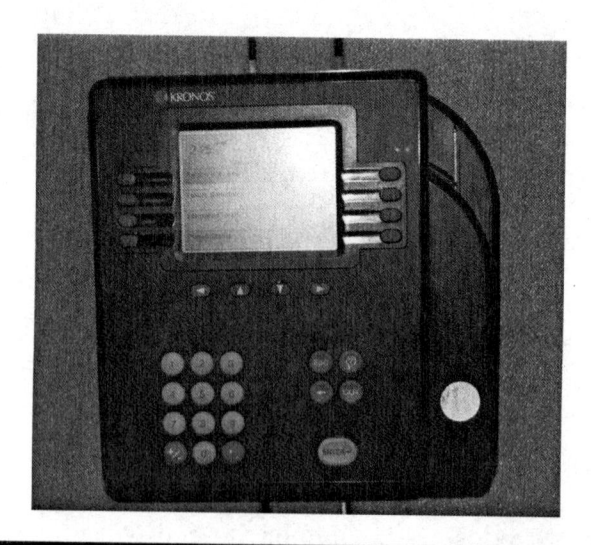

Figure 7.1 Whether a proximity card or a magnetic stripe card, a card reader is one of the basic components of an access control system. (Photo by Daniel J. Benny.)

is given general access to the perimeter doors during normal working hours. It makes no sense to give an employee unlimited access to the facility because the idea is to control access. Anything outside normal business hours will need to be authorized by management. Any areas that are deemed sensitive or secured will need to be authorized by the manager of those areas within the facility.

In general, an ACS compares an individual's badge against a verified database. If authenticated, the ACS sends output signals that allow authorized personnel to pass through controlled areas such as a gate or door. The system has the capability of logging and archiving entry attempts (authorized and unauthorized).

However, in areas where there is a large influx of personnel, such as a lobby entry, it is best to use a combination of ACS and guards. This complements an automated system; guards can detect a large number of people during shift change and adjust with additional personnel for visual identification. Guards can also verify authorized personnel against the badge system and they can react in the event of an emergency or a malfunction of the system. Automated systems cannot account for these types of events. In the event of a fire, automated systems cannot handle the panic that normally ensues when the alarms go off. Traffic is processed at the same pace.

Anti-Passback

In high-security areas, a card reader is utilized on both entry and egress sides of the door. This keeps a record of who went in and out. Anti-passback is a strategy where a person must present a credential to enter an area or facility and then again use the credential to "badge out." This makes it possible to know how long a person is in an area and to know who is in the area at any given time. This requirement also has the advantage of instant personnel accountability during an emergency or hazardous event.

Anti-passback programming prevents users from giving their cards or PIN number to someone else to gain access to the restricted area. In a rigid anti-passback configuration, a credential or badge is used to enter an area and that same credential must be used to exit. If a credential holder fails to "badge-out" properly, entrance into the secured area can be denied.

Card Types

Magnetic stripe (mag stripe) cards consist of a magnetically sensitive strip fused onto the surface of a PVC material, typical to a credit card. A magnetic stripe card is read by swiping it through a reader or by inserting it into a position in a slot. This style of card is old technology; it may be physically damaged by misuse and its data can be affected by magnetic fields.

Proximity card (prox cards) uses embedded antenna wires connected to a chip within the card. The chip is encoded with the unique card identification. Distances at which proximity cards can be read vary by manufacturer and installation. Readers can require the card to be placed within a fraction of an inch from the reader to 6 inches away. This will then authenticate the card and will release the magnetic lock on the door.

Smart cards are credential cards with a microchip embedded in them. Smart cards can store enormous amounts of data, such as access transactions, licenses held by individuals, qualifications, safety training, security access levels, and biometric templates. This card can double as an access card for doors and be used as an authenticator for a computer. The federal government (HSPD-12) has mandated smart cards to provide personal identity verification (PIV) to verify the identity of every employee and contractor in order to improve data security. The card will be used for identification, as well as for facility and data access.

Additional security measures can be employed using keypads with PIN codes. Coded devices use a series of assigned numbers commonly referred to as a PIN. This series of numbers is entered into a keypad and is matched to the numbers stored in the ACS. This provides additional security because if a badge is lost or stolen, it will not activate a control area without the proper PIN number—similarly to an ATM bank card.

Access Control Head End

The application software housed in the CPU is the physical intelligent controller where all access control systems are activity monitored, recorded into history, and commanded and controlled by the operator. Current state-of-the-art access systems such as Lenel OnGuard or Software House CCURE systems, allow each local security panel to hold the system logic for its associated devices. The CPU retains the system-specific programming to allow entry (access) for authorized personnel and deny access to unauthorized personnel.

Communications failure between the CPU and the local access control panels could result in new users not being permitted entry. However, the system is set so that the panel will recognize personnel already installed and will grant access to an authorized badge holder.

These systems have advances that can integrate with CCTV and provide instant visual recognition along with visual alarm activation in order to provide the security console operator visual information before dispatching a security response team.

Receptionist

Receptionists give visitors their first impression of the organization. While on duty, it is necessary for them to maintain a professional appearance and manner. Friendly, positive, and polite attitudes are a requirement for this position.

Depending on your organizational structure, a receptionist may or may not be security personnel. Receptionists need to understand all security procedures and be aware of everything that occurs in the lobby. Keeping a clear view of the surroundings is very important and all suspicious activity should be reported to supervision.

Visitors are guests, customers, or vendors or in some way have business dealings with the company. As such they should be greeted by a knowledgeable receptionist who gives the visitor his or her first impression of the organization. There should be some type of controlled waiting area within the lobby so that the receptionist can keep an eye on the visitor and can direct the employee to him or her in the event that they have not met previously.

Escort and Visitor Control

All visitors entering the facility should sign in and out on a visitors' log to maintain accountability of who is in the facility, the time frame of the visit, whom they visited, and, in the case of an emergency, to have accountability of everyone for safety purposes.

Visitors are given temporary badges, but this badge does not double as an access card. The temporary badge will be issued at an entry control point only after the visitor identifies the purpose of the visit and receives approval by the employee being visited. In some organizations, only certain employees may approve visitor access along with the day and time of the visit. In many operations, the visitor is escorted at all times while inside the facility.

When the visitor arrives, he or she will present a form of photo identification, such as a driver's license, to the receptionist for verification. Some visitor badges are constructed of paper and may have a feature that causes a void line to appear after a preset time period. Typically, the pass is dated and issued for a set period, usually 1 day. In most cases a visitor will wear a conspicuous badge that identifies him or her as a visitor and clearly indicates whether an escort is required (often done with color-coded badges). If an escort is required, the assigned person should be identified by name and held responsible for the visitor at all times while the visitor is on the premises.[1]

A visitor management system can be a pen and paper system that records basic information about visitors to the facility. Typical information found in an entry includes the visitor's name, reason for the visit or who the visitor is visiting, date of the visit, and the check-in and check-out times.

Other types of visitor management systems use a computer-based system or specific visitor software product. They can be either manually inserted into the system by the receptionist or, on a higher end visitor management system, the visitor provides the receptionist with identification, such as a driver's license or a government or military ID. The receptionist then swipes the person's identification through a reader. The system automatically populates the database with ID information and recognizes whether the ID is properly formatted or false. The receptionist who is registering the guest identifies the group to which the person belongs—guest, client, vendor, or contractor. Then the badge is printed.

It is best for the employee to come to the lobby area to greet the visitor personally. This is more than a common courtesy because it provides the necessary security in proper identification and escorting and controlling the movement of the visitor. Some companies initiate a sound security practice by properly identifying the visitor and signing him or her into a visitor management system—but the visitor is then allowed to wander the halls of the company trying to locate the employee to be visited. In this respect, this does not make good business or security sense.

Once the visitor has completed the visit or meeting, the employee or escort brings him or her down to the lobby, gives the visitor badge back to the receptionist, and then logs the departure time into the system. If, at the end of the day, a badge has not been returned, the receptionist will contact the employee and inquire as to the whereabouts of the visitor or the visitor badge. It is also recommended that no visitor badges be allowed outside the controlled area so that anyone attempting to reenter will pass by the receptionist and will be stopped. If you allow visitors to leave the facility with the temporary visitor badge, what guarantee do you have that the badge is being carried by the right person unless you repeat the sign-in process? It is best to have the visitor return the badge to the receptionist if he or she leaves the protected area so that the receptionists can control this area of responsibility.

Other functions can be enabled with a reusable visitor badge that has been programmed into the access control system from the receptionist desk. In this scenario, even after an employee who is escorting the visitor proceeds into the facility, the visitor will still be required to use the visitor badge to gain entry. There will then be an electronic time stamp of when this visitor went into the facility, and if the badge is not returned, it can be deactivated and cannot be used for access. It is then just a worthless piece of plastic.

For facilities that have large meetings with outside guests, it is recommended that the initiating internal group that is putting on the meeting prepare a visitor log with the individuals' names and organizations. This will assist the receptionist in preparing badges in advance and having them ready for the meeting day. The same principle of escorting individuals applies; from a security standpoint, one escort employee can properly control five visitors.

Guards

According to the Department of Labor, "Security guards, also called security officers, patrol and inspect property to protect against fire, theft, vandalism, terrorism, and illegal activity. These workers protect their employer's investment, enforce laws on the property, and deter criminal activity and other problems."[2]

Security officers are the physical presence and the deterrence to unauthorized entry into the facility along with being the response force to an alarm activation. Despite all the alarm technology, it still requires human intervention to respond to an alarm, make contact with an intruder, interact with employees, and provide first-aid when necessary.

While many companies are attempting to cut costs by reducing the amount of physical security presence with electronic means, the need for physical security still exists. Physical security is still needed to control entry points, even with electronic entry control systems. The goals of entry control drive the use of both electronic and physical guards for entry control points.

The use of physical guards for entry control depends upon the level of protection required for a defined asset. Physical guards have the advantage of being able to detect and assess an event in split-second timing. For example, a guard at a vehicle entry control point on the outer perimeter of your installation can respond with immediate notification if a vehicle breaks though the perimeter. In this situation, the response time is faster and notification can be made to a response team faster than if the event is detected via an unmanned sensor.

Guards are ideal in situations where entry control points have extra traffic that needs to be accounted for. This could be during shift changes, fire alarms, or evacuations.[3] They are also used for verifying personnel that are entering an area by using visual identification. This includes using photo image badges, monitoring exchange badge systems, and reviewing facial recognition software. There is a unique psychological effect in the presence of a well-trained and professional-looking security guard compared to the presence of a CCTV camera.

Security officers are required to conduct foot patrols of building interiors, exteriors, and parking areas. Some officers are assigned a fixed or stationary position at entrances and other designated areas in order to prevent unauthorized entrance or the introduction of prohibited items. Another security officer responsibility is to control access into the facility by checking employee identification badges, issuing temporary badges, and registering visitors. Officers are required to respond to fire, security, and medical emergencies, and to render assistance when needed as well as submit written or verbal reports regarding significant events to security management. They also escort designated visitors, usually construction or maintenance contractors, who require nonbusiness-hour access to facilities or access to areas where classified or proprietary information is accessible. They must report potentially hazardous conditions and items in need of repair immediately, including inoperative lights, leaky sprinkler heads, leaky faucets, toilet stoppages, broken or slippery floor surfaces, trip hazards, etc.

While guards are a valid security measure, they can also be the weakest link in a physical protection system. Human nature must be factored into the equation. Security is only as reliable as the weakest link. If an alarm console system is poorly designed, the effectiveness of the operator's ability to monitor and assess events is degraded. Security guards within the private sector receive minimal to no training, receive no benefits, and have little vested in the company that they are hired to protect.[4] This creates scenarios in which a guard may not look as hard at an identification badge, overlook an event due to a distraction, or just not care. This puts the pressure on security operations to compensate for the guard in detecting, delaying, and responding to alarms. Automated systems are never the best option. If a system is fully automated and it is bypassed, the system fails.

Turnstiles and Mantraps

A common and frustrating loophole in an otherwise secure access control system can be the ability of an unauthorized person to follow through a checkpoint behind an authorized person. This is called "piggybacking" or "tailgating."

The traditional solution is an airlock-style arrangement, called a "mantrap," in which a person opens one door and waits for it to close before the next door will open. A footstep-detecting floor can be added to confirm that only one person is passing through. A correctly constructed mantrap or portal will provide for tailgate detection while it allows roller luggage, briefcases, and other large packages to pass without causing nuisance alarms. People attempting to enter side by side are detected by an optional overhead sensing array. The mantrap controller prevents entry into secured areas if unauthorized access is attempted (Figure 7.2).

Figure 7.2 A mantrap. (Photo by Daniel J. Benny.)

(a)

Figure 7.3(a)   (a–c): A turnstile can be used as a supplemental control to assist a guard or receptionist while controlling access into a protected area. (continued)

Another system that is available is a turnstile, which can be used as a supplemental control to assist a guard or receptionist while controlling access into a protected area. Anyone who has gone to a sporting event has gone through a turnstile. In this approach, the individual's badge is used to control the turnstile arm and allow access into the facility (Figure 7.3a–c).

A higher end turnstile is an optical turnstile, which is designed to provide secure access control in the lobby of a busy building. This system is designed as a set of parallel pedestals that form lanes that allow entry or exit. Each barrier is equipped with photoelectric beams, guard arms, and a logic board.

To gain access to the interior of the building, an authorized person uses his or her access card at the optical turnstile. When the access card is verified, the guard arm is dropped, the photoelectric beam is temporarily shut off, and the cardholder passes without creating an alarm.

(b)

Figure 7.3(b) (continued) (a–c): A turnstile can be used as a supplemental control to assist a guard or receptionist while controlling access into a protected area. (continued)

The concept behind these options is to create a secure perimeter just inside the building to ensure that only authorized people proceed further into the building, thereby creating the secure working environment.

(c)

Figure 7.3(c) (continued) (a–c): A turnstile can be used as a supplemental control to assist a guard or receptionist while controlling access into a protected area.

References

1. Fennelly, L. 2004. *Effective physical security.* 3rd ed., 195. New York: Elsevier.
2. http://stats.bls.gov/OCO/OCOS159.HTM
3. Garcia, M. L. 2006. *Vulnerability assessment of physical protection systems.* Burlington, MA: Elsevier Butterworth-Heinemann.
4. Fischer, R., E. Halibozek, and G. Green. 2008. *Introduction to security.* New York: Elsevier.

Chapter 8
Physical Protection Systems

Paul R. Baker

Doors

Door assemblies include the door, its frame, and anchorage to the building. As part of a balanced design approach, exterior doors should be designed to fit snugly in the doorframe, preventing crevices and gaps, which also helps prevent many simple methods of gaining illegal entry. The doorframe and locks must be as secure as the door in order to provide good protection.

Perimeter doors should consist of hollow steel doors or steel-clad doors with steel frames. Ensure that the strength of the latch and frame anchor equals that of the door and frame. Permit normal entry/egress through a limited number of doors, if possible, while accommodating emergency egress. Ensure that exterior doors into inhabited areas open outward. Locate hinges on the interior of restricted areas. Use exterior security hinges on doors opening outward to reduce their vulnerability.

If perimeter doors are made of glass, make sure that the material is constructed of a laminate material or stronger. Ensure that glass doors only allow access into a public or lobby area of the facility. High-security doors will then need to be established within the lobby area, where access will be controlled.

All doors that are installed for sensitive areas such as telephone closets, network rooms, or any area that has access control will require the door to have an automatic door-closing device.

Door Locks

Electric Locks

The electric lock is a very secure method to control a door. An electric lock actuates the door bolt. For very secure applications, dual locks can be used. In some cases, power is applied to engage the handle, so the user can retract the bolt instead of the electric lock door operator actually retracting the bolt. Most electric locks can have built-in position switches and request-to-exit hardware. Although offering a high security level, electric locks are expensive. A special door hinge that can accommodate a wiring harness and internal hardware to the door is required. For retrofit applications, electric locks usually require purchase of a new door.

Electric Strikes

The difference between an electric strike and an electric lock is in the mechanism that is activated at the door. In an electric-lock door, the bolt is moved. In an electric-strike door, the bolt remains stationary and the strike is retracted. As in electric locks, electric strikes can be configured for fail-safe or fail-secure operation. The logic is the same.

In fail-safe configuration, the strike retracts when de-energized on loss of power. This allows the door to be opened from the public side. In fail-secure configuration, the strike remains in place, causing the door to be locked from the public side and requiring manual key entry to unlock the door from the public side. Again, as with electric locks, unimpeded access is allowed in the direction of egress by manual activation of the door handle/lever when exiting from the secure side. For retrofit situations, electric strikes rarely require door replacement and can often be done without replacing the doorframe.

Magnetic Locks

The magnetic lock is popular because it can be easily retrofitted to existing doors (Figure 8.1). The magnetic lock is surface mounted to the door and doorframe. Power is applied to magnets continuously to hold the door closed. Magnetic locks are normally fail-safe.

Magnetic locks do have a security disadvantage. In requirements for life safety codes, doors equipped with magnetic locks are required to have one

Figure 8.1 Electromagnetic door contact sensor. (Photo by Daniel J. Benny.)

manual device (emergency manual override button) and an automatic sensor (typically, a passive infrared sensor, PIR, or request-to-exit, REX) to override the door lock signal when someone approaches the door in the exit direction. All locks are controlled by a card reader that, when activated, will release the secured-side portion of the door and allow entry into the facility.

While enhancing overall building safety, the addition of these extra devices allows possible compromise of the door lock. In the scenario where a REX device is used with magnetic locks, it not only turns off the alarm when the individual exits but also deactivates the locking device. This can be a problem if an adversary can get something through or under the door to cause the REX to release the magnetic lock.

Windows

Because of the ease with which most windows can be entered or glass broken, they are targets for most intruders; therefore, they need to be addressed as a potential vulnerability in the facility defenses. A standard home's installed glass windows can be shattered relatively easily when hit with force. The glass not only will break but also will leave sharp fragments that can cause severe lacerations.

Window systems such as glazing, frames, and anchorage to supporting walls on the exterior facade of a building should be used to mitigate the hazardous effects of flying glass during an explosion event. In an effort to protect occupants, the security professional should integrate the features of the glass, the connection of the glass to the frame, and anchoring of the frame to the building structure to achieve a balanced installation.

It is recommended that windows should not be placed adjacent to doors because, if a window is broken, the door can be reached and unlocked. Consider using laminated glass in place of conventional glass and placing window guards, such as grilles, screens, or meshwork, across window openings to protect against covert entry. Windows on the ground level should not have the ability to open and should be protected with bars and alarm systems. The alarms available for a window include a magnetic switch, which when the magnets are separated (as when the window is opened), causes an alarm to sound. It is recommended that windows up to the fourth floor should have this protection installed. Also, consider using steel window frames securely fastened or cement grouted into the surrounding structure.

Types of Glass

Tempered glass is similar to the glass installed in car windshields. It will resist breakage and will disintegrate into small cubes of crystals with no sharp edges. Tempered glass is used in entrance doors and adjacent panels.

Wired glass provides resistance to impact from blunt objects. The wire mesh is embedded into the glass, thereby providing limited protection.

Laminated glass is recommended for installation in street-level windows, doorways, and other access areas. It is made from two sheets of ordinary glass bonded to a middle layer of resilient plastic. When it is struck, it may crack but the pieces of glass tend to stick to the plastic inner material.

Bullet-resistant (BR) glass is typically installed in banks and high-risk areas (Figure 8.2). There are different layers of BR glass with the standard being 1¼-inch thick, which provides protection from a .9mm round.

Glass-Break Sensors

Glass-break sensors are a good intrusion detection device for buildings with a lot of glass windows and doors with glass panes. Glass as an exterior protection barrier can be easily defeated. Windows can be quickly and easily broken.

Figure 8.2 Lexgard® bullet-resistant glass that has stopped a .44 caliber bullet. (Photo by Daniel J. Benny.)

There are several basic types of glass-break sensors. Acoustic sensors listen for an acoustic sound wave that matches the frequency of broken glass and shock sensors feel the shock wave when glass is broken. The use of dual-technology glass-break sensors (acoustic and shock wave) is most effective because if only acoustic is used and an employee pulls the window blinds up, this can set off a false alarm. However, if the system is set to a dual alarm, both acoustic and shock will need to be activated before an alarm is triggered. There is not a significant price difference between a simple acoustic sensor and combination sensors (acoustic and shock). For the nominal component price increase, which is a fraction of the total installed cost, the increased capability justifies the higher price tag.

Interior Intrusion Detection Systems

An intrusion detection system is designed to provide notice of someone entering a protected area through a system of sensors that send a notification to the computer base's monitoring stations or to a local sound-producing device. The intrusion detection system can be a proprietary central station in which it is monitored by an organization's security department or by use of a contract central station. The contract central station is a contract security monitoring service not located at or associated with the facility being protected.

Within the facility, it is still necessary to maintain levels of security. Some operations feel that security should be like an M&M's® candy that is hard on the outside and soft on the inside—meaning that all the security is on the perimeter but when you gain access, you should have free reign throughout the facility. This approach is not conducive to a protection-in-depth security system. The layered approach provides for additional security measures while inside the perimeter of the facility. Specifically, not all employees need access to the sensitive area, such as the phone closets, or need access to the data center. It is not practical or economical to have guards stationed at every security point within the facility; however, an access control system can provide the necessary security controls throughout the building.

A card reader can control access into a specific room. This can be controlled through the access control software, which will be maintained within the security control center. If the individual has access to the room, the employee will place his badge up to the reader and it will release the electric lock and allow entry.

Other elements necessary for this control of interior access are discussed next.

Balanced Magnetic Switch (BMS)

This device uses a magnetic field or mechanical contact to determine if an alarm signal is initiated. One magnet will be attached to the door and the other to the frame; when the door is opened, the field is broken. A BMS differs from standard magnetic status switches in that a BMS incorporates two aligned magnets with an associated reed switch. If an external magnet is applied to the switch area, it upsets the balanced magnetic field such that an alarm signal is received. Standard magnetic switches can be defeated by holding a magnet near the switch. Mechanical contacts can be defeated by holding the contact in the closed position with a piece of metal or taping contacts closed. Balanced magnetic switches are not susceptible to external magnetic fields and will generate an alarm if tampering occurs. These switches are used on doors and windows (Figure 8.3).

Motion-Activated Cameras

A fixed camera with a video motion feature can be used as an interior intrusion point sensor. In this application, the camera can be directed at an entry door and will send an alarm signal when an intruder enters the field

Figure 8.3 A balanced magnetic switch (BMS), top right, used on doors and windows, uses a magnetic field or mechanical contact to determine if an alarm signal is initiated. (Photo by Daniel J. Benny.)

of view. This device has the added advantage of providing a video image of the event, which can alert the security officer monitoring the camera so that he or she can make a determination of the need to dispatch a security force.

Typically, one camera can be associated with several doors along a hallway. If a door is forced open, the alarm will trigger the camera to begin recording and can give the monitoring officer a video view starting 1 minute before the alarm was tripped, so as to allow the operator all the possible information before dispatching a security response. This system uses technology to supplement the guard force. It can activate upon motion and can give a control center operator a detailed video of actual events during alarm activation.

Acoustic Sensors

This device uses passive listening devices to monitor building spaces. An application is an administrative building that is normally only occupied in daylight working hours. Typically, the acoustic sensing system is tied into a password-protected building entry control system, which is monitored by a central security monitoring station. When someone has logged into the building with a proper password, the acoustic sensors are disabled. When the building is secured and unoccupied, the acoustic sensors are activated. After business hours, intruders make noise that is picked up by the acoustic array and an alarm signal is generated.

The downside is the false alarm rate from picking up noises such as air conditioning and telephone ringers. This product must be deployed in an area that will not have any noise. Acoustic sensors act as a detection means for stay-behind covert intruders. One way to use the system is as a monitoring device: When it goes into alarm, the system will open up an intercom and the monitoring officer can listen to the area. If no intruder is heard, then the alarm is cancelled.

Infrared Linear Beam Sensors

Most of us think of this device from James Bond and Mission Impossible movies, where the enduring image of secret agents and bank robbers donning their special goggles to avoid triggering an active infrared beam is recalled. This is the device most of us have on our garage doors. A focused infrared (IR) light beam is projected from an emitter and bounced off a reflector that is placed at the other side of the detection area. A retroreflective photoelectric beam sensor built into the emitter detects when the infrared beam is broken by the passing of a person or the presence of an object in the path of the infrared beam. If the beam is broken, the door will stop or the light will come on. This device can also be used to notify security of individuals in hallways late at night, when security is typically at its reduced coverage.

Passive Infrared (PIR) Sensors

A PIR sensor (Figure 8.4) is one of the most common interior volumetric intrusion detection sensors. Because there is no beam, it is called passive. A PIR picks up heat signatures (infrared emissions) from intruders by comparing infrared receptions to typical background infrared levels. Infrared radiation exists in the electromagnetic spectrum at a wavelength that is longer

Figure 8.4   A passive infrared (PIR) sensor is one of the most common interior volumetric intrusion detection sensors. Because there is no beam, it is called passive.

than visible light. It cannot be seen but it can be detected. Objects that generate heat, including animals and the human body, also generate infrared radiation. The PIR is set to determine a change in temperature, whether warmer or colder; it distinguishes an object that is different from the environment that it is set in. Typically, activation differentials are 3°F. These devices work best in a stable, environmentally controlled space.

A PIR is a motion detector and will not activate for a person who is standing still because the electronics package attached to the sensor is looking for a fairly **rapid change** in the amount of infrared energy it is seeing. When a person walks by, the amount of infrared energy in the field of view changes rapidly and is easily detected. You do not want the sensor detecting slower changes, like the sidewalk cooling off at night.

PIRs come in devices that project out at a 45° angle and can pick up objects 8 to 15 meters away. There are also 360° PIRs, which can be used in a secured room, so that when there is entry the PIR will activate. These motion detection devices can also be programmed into an alarm keypad located within the protected space. When motion is detected, it can be programmed to wait for a prescribed time while the individual swipes a badge or enters pass code information into the keypad. If identification is successful, the PIR does not send an intruder notification to the central station.

PIRs not only are a security application, but also are often used as an automatic REX device for magnetically locked doors. In this application, the REX acts as the automatic sensor for detecting an approaching person in the exit direction for magnetically locked doors and deactivates the alarm.

Dual-Technology Sensors

These provide a commonsense approach for the reduction of false-alarm rates. For example, this technology uses a combination of microwave and PIR sensor circuitry within one housing. An alarm condition is only generated if both the microwave and the PIR sensor detect an intruder. Since two independent means of detection are involved, false-alarm rates are reduced when configured into this setting. Integrated, redundant devices must react at the same time to cause an alarm. More and more devices are coming with dual technology that will reduce the need for multiple devices and significantly reduce the false-alarm rates.

Almost all protection systems include intrusion detection and fire safety into one integrated system. The fire protection system can be activated manually by use of a pull station should one smell or see smoke and fire. The pull station will activate the audible and visual strobe and fire protection enunciators in the building, and notify the central station and/or emergency dispatch for the fire department. In addition to the manual pull station, a fire protection sensor can be placed in the protected facility that will send an automatic signal to the central station and/or emergency dispatch for the fire department and activate a set of the audible and visual strobe fire protection enunciators. The following fire protection sensors will be utilized:

- Dual chamber smoke detector. This sensor will provide early detection of smoke. It is used primarily for the protection of life, but early detection of a fire can also save property.
- Rate-of-rise heat detector. This sensor is used in an area where a smoke detector cannot be used because the normal activity in those areas would set off a smoke detector. This would include bathrooms, cooking areas, and workshops. The rate-of-rise heat detector will sense a rapid increase of the heat in an area due to a fire and will then activate the alarm system.
- Natural gas or carbon monoxide detectors. These two types of sensors are used to provide early warning for evacuation due to deadly gases that may build up in a facility.
- Water flow. For facilities that have fire protection sprinkler systems, this sensor will detect the drop in water pressure when the sprinkler is activated during a fire. This will result in an alarm being activated.

Perimeter Intrusion Detection Systems

Depending on the extent of security required to protect the facility, exterior or perimeter sensors will alert you to any intruders attempting to gain access across open space or attempting to breach a fence line. These may provide security ample opportunity to evaluate and intercept any threat.

In general, open terrain sensors work best on flat, cleared areas. Heavily or irregularly contoured areas are not conducive to open terrain sensing systems. Open terrain sensors include infrared, microwave systems, combination (dual technology), vibration sensors, and newly emerging video content analysis and motion path analysis (e.g., CCTV) systems.

Infrared Sensors

Passive infrared sensors are designed for human body detection, so they are great for detecting when someone is approaching.

> Passive-infrared sensors detect the heat emitted by animate forms. Because all living things emit heat, a system of recording measurable changes in a specific area provides a means of detecting unauthorized intrusions. When the unit registers changes in temperature in its area of detection, it relays the information to a processor which measures the change according to detection parameters. If the change falls outside the parameters, the processor sends a signal to the unit's alarm.[1]

Active infrared sensors send an infrared signal via a transmitter to a receiver. Interruption of the normal IR signal indicates that an intruder or object has blocked the path. The beam can be narrow in focus, but should be projected over a cleared path.

Microwave Systems

Microwave sensors (Figure 8.5) come in two configurations: bistatic and monostatic. Both types of sensors operate by radiating a controlled pattern of microwave energy into the protected area. The transmitted microwave signal is received, and a base level "no intrusion" signal is established. Motion by an intruder causes the received signal to be altered, setting off an alarm. Microwave signals pass through concrete and steel and must be applied

Figure 8.5 A microwave sensor on the White House lawn. (Photo by Daniel J. Benny.)

with care if roadways or adjacent buildings are near the area of coverage. Otherwise, nuisance alarms may occur due to reflected microwave patterns.

A bistatic sensor sends an invisible volumetric detection field that fills the space between a transmitter and receiver. Monostatic microwave sensors use a single sensing unit that incorporates both transmitting and receiving functions. It generates a beam radiated from the transceiver to create a well-controlled, three-dimensional volumetric detection pattern with adjustable range. Many monostatic microwave sensors feature a cutoff circuit, which allows the sensor to be tuned to cover only the area within a selected region. This helps to reduce nuisance alarms.

Coaxial Strain-Sensitive Cable

These systems use a coaxial cable that transmits an electric field and is woven through the fabric of the fence. As the cable moves due to strain on the fence fabric caused by climbing or cutting, changes in the electric field are detected within the cable, and an alarm condition occurs.

Coaxial strain-sensing systems are readily available and are highly tunable to adjust for field conditions due to weather and climate characteristics. Some coaxial cable systems are susceptible to electromagnetic interference and radio frequency (RF) interference.

Time Domain Reflectometry (TDR) Systems

TDR systems send induced RF signals down a cable attached to the fence fabric. Intruders climbing or flexing a fence create a signal path flaw that can be converted to an alarm signal. When the conductor cable is bent or flexed, a part of the signal returns to the origination point. This reflected signal can be converted to an intrusion point by computing the time it takes for the signal to travel to the intrusion point and return. The cable can be provided in armored cable, which requires more than a bolt cutter to sever the sensing cable. These systems require their own processor unit and can be configured in a closed loop so that if the cable is cut, it can be detected by the other return path.

Video Content Analysis and Motion Path Analysis

The newest technology for intrusion detection is sophisticated software analysis of camera images such as video content analysis and motion path analysis. CCTV camera systems are increasingly being used as intrusion detection systems. Application of complex algorithms to digital CCTV camera images allows these systems to detect intruders. The software programming is smart enough to detect pixel changes and differentiate and filter out normal video events (leaves blowing, snow falling) from true alarm events. The application of software rules can further evolve to differentiate between a rabbit hopping across a parking lot and a person trespassing through the parking lot, which needs to be addressed.

The application of complex software algorithms to CCTV digital images takes on the aspect of an artificial camera, whereby the camera and processors become "smart video" and start to emulate a human operator. The differences between a smart camera and a human operator are principally twofold. It takes a lot of complex software programming and associated rules to give the camera systems the ability to differentiate and assess video events compared to the processing ability of the human mind.

With more and more project applications, the gap is closing as the camera systems come closer to emulating the capabilities of a fully alert, very motivated, intelligent security guard fresh into the shift. The advantage of video content analysis and motion path analysis is that the camera systems do not get tired. Studies have demonstrated that after 20 minutes, the ability of a guard to discern an abnormal event is severely degraded. Video content analysis systems do not suffer fatigue and remain "alert" after monitoring hundreds of video events during a shift. Video content analysis systems can monitor more cameras, more effectively, with fewer operators at a reduced cost. This will allow fewer dispatch center/command center staff while letting technology assist with the human factor.

References

1. http://www.homesecurityguru.com/passive-infrared-pir-sensors

Chapter 9

CCTV and IP Video

Steve Surfaro

The Essential Guide to Video Surveillance

There are few subjects that dominate publications, education sessions, negotiations, and actual deployments as much as video surveillance does in the physical security industry. Video surveillance has made a huge impact in many other industries, as well, and is often a compliance requirement, such as with gaming, in order to obtain a state gaming license. SAS70 and PCI compliance drive the specification and deployment of video surveillance in data centers or wherever there are customer credit card transactions. HIPAA requirements do not directly require video surveillance to protect, for example, a patient record room, but recorded video systems are common in these environments as well.

Video surveillance systems are often more specifically called analog video surveillance, or IP video surveillance, to indicate their use of conventional or network-based connectivity. The older "video surveillance" term generally indicates analog video surveillance, but has been used generally to refer to video of all types, except those supported by cloud computing and, more specifically, hosted video systems supported by a software as a service (SaaS) deployment.

When considering the design and selection of a video surveillance system, the physical security designer, user, or integrator needs to consider the individual needs of each use case and market with which he or she is working. The following are intended as examples only and each market's own requirements should be verified on a project-by-project basis. The term "use case" is pretty much the same thing as an "application," but we are trying to avoid confusion with terminology identifying actual software, so we call the different ways we deploy video surveillance "use cases."

Video Surveillance Use Cases

The following is a summary of the most popular use cases, followed by most popular function:

- In-car and transit video surveillance
  - Provide most usable wide view surveillance products for identification
  - Provide rugged, removable digital media for video storage
- Retail video and loss prevention
  - Design POS transaction data ("electronic journal") with digital video images of the events for simultaneous viewing and proof
  - Use input from one or multiple POS or ATM machines via RS-232 interface or Ethernet
  - Provide simultaneous viewing of time-linked transactions and multiple video cameras
  - Transaction searching and "go to" next or previous match
  - Print transaction reports of retail shrinkage events with linked video images
- Campus energy environment
  - Provide products that may be initially more expensive, but have longer life and extremely low maintenance
- Retail and corporate banking
  - Application driven by flexibility, low qualified cost, and the ability to produce excellent images for forensic purposes
- Urban surveillance
  - Provide low-light capability for all outdoor public video surveillance devices
  - Provide cameras capable of producing high-resolution video images and the addition of a video analytics subsystem where required
  - Provide compatibility with fiber optic or wireless transport systems
- Public arenas
  - Provide cameras capable of producing high-resolution video images and the addition of a video analytics subsystem where required

- Provide low-light and IR-compatible cameras where required
- Provide control system with minimal camera control and switching delay
- Provide ease of expansion to accommodate specialized surveillance of visiting dignitaries
- Residential
  - Low cost, WAN (wireless area network) friendly, and periodic, rather than continuous, duty mark this category's requirements
- Gaming
  - Continuous duty
  - High-resolution, especially in high ceiling areas
  - Accurate color reproduction
  - Provide control system with minimal camera control and switching delay
  - Provide video management or video recording systems with instant replay capability for multiple operators and multiple cameras
  - Redundancy of all operation and equipment capabilities required, especially in "high stakes" areas

Video Surveillance Impacts Many Other Industries as Its Use Expands

To reduce the "shrink" or loss in an operation, retail markets use video surveillance—quite often through a specialized subset of video surveillance technology, video analytics, and an analysis "snapshot" in time compared with video content analysis (VCA), which analyzes video data by single or multiple criteria and then delivers a search result. This is not to be confused with a newer technology, known as video summarization or synopsis, that artfully condenses an entire day of video to a matter of minutes—a very useful tool!

Transportation video surveillance gives us early warning of traffic events, persons in need of assistance, and simply making normal cyclic traffic flow in a major city easier.

In difficult economic times, mass transit ridership increases, and the opportunity for criminal behavior does as well. Video surveillance creates a "safe zone" after-hours waiting area and offers prevention of undesirable events. Mass transit operators also use video verification to locate transit vehicles in tunnels. Finally, in-car video surveillance gives an opportunity to replay video content from the time leading up to and after an event, in addition to real-time video observation.

Casino surveillance operators are looking at a system where cameras outnumber the security system in the same facility by a factor of 10–100:1. The purpose of the system is primarily to observe players in real time, but high-resolution clips can be played back at any time.

Public safety professionals and first responders use video live feeds from HDTV cameras to assess an event in progress and determine manpower response. Emergency medial technicians use remotely transmitted video to confirm a diagnosis with a health professional remotely located at a hospital taking in the patient. HDTV cameras record an operation in progress for current observation and future education and distance medical learning.

Perimeter surveillance cameras located at an airport acquire an image of and analyze an unattended vehicle and notify the airport law enforcement force. Although these cameras were not being observed, video analytics within the camera and at a video management system provided a real-time notification of a possible threat.

In each of these use cases video surveillance was used in real-time observation, forensic review, and recognition—the three major use classifications of a video surveillance system. Classifying our system helps us use it better and communicate its primary function as a tool for many professionals—not only physical security.

Video Surveillance System Classifications

Classifying video surveillance systems by function allows us to think of benefit first and technology second, as these video surveillance systems are simply tools for us to use. If the solution has interesting features but is not useful for the actual use case, the designer has not adequately identified the video surveillance user's needs.

The three primary classifications of video surveillance systems are observation, forensic review, and recognition. Systems meant for observation do not have as high resolution requirements as recognition, but require high frame (refresh) rates. Systems used primarily for review after an incident occurs must have excellent coverage and frame rate high enough to capture an event. The wonderfully wide (16:9) aspect ratio of HDTV is often an excellent solution for these types of systems as they give the user the best opportunity at providing evidence. Recognition-based solutions or systems that analyze video and provide a result, like license plate capture or recognition, require the highest resolution or amount of pixels on target. A nice way to deploy systems like these and conserve storage or bandwidth is to provide a trigger to activate the recognition-based recording. Install a vehicle loop detector and activate a camera that performs a license plate recognition function.

Analog Video Systems Overview

If a system is primarily used for observation or surveillance, design emphasis placed on the viewing refresh rate will be a benefit. Minimizing the control latency (both camera control delay and switching delay) is extremely important and is one benefit that an analog system offers. Most IP video systems can provide smooth, lag-free control of positionable or PTZ (pan-tilt-zoom) cameras, but this needs to be verified under a variety of network usage scenarios (Figure 9.1).

One of the most challenging things for a surveillance operator is trying to follow a subject using a PTZ camera on a network-based video system that was not designed with a fast control response in mind. Today's analog-based systems are, in some cases, more costly to deploy because they use proprietary cabling infrastructure (such as coaxial cable), but offer a way to minimize control signal delays, even if a comparable network-based system is used for transport, or dedicated communications are used.

Figure 9.1 Two pan-tilt-zoom cameras and a domed camera in an outdoor application.

IP Video Systems Overview

What is video over IP? Video over IP is best defined as the deployment of video information deployed over a network that conforms to the OSI layer model (Figure 9.2). This includes support of cameras and encoders transmitting using various protocols including TCP/IP, UDP, and FTP. Just as e-mails, streaming video, data, and all web traffic travels via coaxial cabling, so too can video data and information from surveillance systems, access control systems, and other electronic physical security devices and systems.

Devices that stream video over IP networks transmit frames and packets of video data to a single location or multiple locations for different purposes. A device like a network video camera or multichannel video encoder can send a video stream to a single network video recorder or video decoder location, or to multiple locations of the same type of equipment.

Why should you choose IP video? In many cases, there is a lower total cost of ownership for life cycle of the system. With a network infrastructure in place, there are lower installation costs when power over Ethernet (PoE, 802.3af) devices are used, eliminating the need for and vulnerability of localized power supplies. An end user can enjoy reduced operational overhead as the IP video devices are accessible at any location on the network. The IP video system devices often require less maintenance and system downtime; however, the designer is cautioned to make sure that support for the shared infrastructure, such as network switches and recording servers, is in place. If not, the IP video system can actually be far more costly to operate and maintain.

Figure 9.2 Video can travel and stream wirelessly and over fiber optic and traditional cables just as e-mail, data, and web traffic can.

The typical IP video system is deployed and expanded much more quickly and easily as the initial infrastructure investment has been installed and future devices are connected to the nearest accessible network access location, which is usually the telecommunications room. When you compare this with installation of point-to-point infrastructure and if the facility in question has changing requirements, the deployment of an IP video system becomes a necessity.

Live or recorded video viewed anywhere, anytime permits a better use of the video system investment. IP video systems are essential parts of public surveillance systems like city center security and mass transit, as there may be an instantaneous requirement to view a single camera by one or many individuals simultaneously.

"Open" infrastructure permits integration of video surveillance devices from different manufacturers; however, the designer is cautioned that challenges similar to the analog video system exist where interoperability is required. Even though the transport is standardized with IP video, each manufacturer uses a special version of a video compression engine in the video source that must be decoded by the recording and control system.

There is an accepted method of exchanging such software development kit (SDK) and application programming interface (API) application programming interface information between manufacturers, and some have even standardized on a single API to ease the burden on the developer. Choosing an IP camera made by a manufacturer having a strong partner program is like buying an investment in a platform. There are many developers that have created unique and creative solutions that can better fit your need. Purchasing an IP camera from an unpopular manufacturer will greatly limit the different recording and monitoring solutions you may use.

Standards of interoperability help the decision process of an IP camera selection, but the designer must be ready to score compliance with required standards of interoperability. Remember that there is no partial compliance; a solution either conforms or it does not!

Assuming that integration is proceeding, it is far easier to link disparate systems to a single platform. The integration of related systems, including access control, intrusion, fire/safety, and communication can provide for beneficial interoperability.

Higher-end imaging like megapixel and progressive scans are available in the IP video system and represent a distinct advantage over their analog counterparts. It is necessary to keep in mind that many of the specialized imaging technologies still remain analog and must be converted to IP video through the use of video servers or encoder devices. Megapixel and progressive scan imaging make it easier for objects and individuals to be identified in recordings than their nearest equivalent analog counterparts, assuming that the same imaging technologies are used in both devices.

Direct attached storage and storage area networks are easily scalable and provide future growth for the IP video platform, as they can be placed and managed at or from any accessible location on the network. With IP video, there is a need to reduce the risk to your facility's infrastructure, but best practices for network security design usually place devices like these well behind corporate firewalls and not generally accessible to the public.

Should the IP video device be publicly accessible, the risks are more of denial of service as the IP video camera will only support a finite number of users directly. There is also the risk of taking control of the device itself and redirecting the video stream elsewhere, in addition to accessing any stored video data within the camera.

It is for this reason that many manufacturers have adopted the use of Port Authentication Protocol (802.1x) to better manage video streaming and accessibility of the recording and monitoring devices that exist directly on a corporate network. The designer is cautioned, however, to model all systems that utilize authentication protocols to assure the level of performance that their user requires.

Image Capture Video Sources—Cameras

When you consider camera technology, functionality, and use, there are several major classifications, including fixed, fixed zoom, PTZ, unitized dome, board cameras, and miniature cameras.

Analog Video Cameras

Today's analog cameras are actually fully digital inside as they contain a DSP (digital signal processor) that allows for various image processing technologies to be applied. Digital signal processing permits the images projected by the lens on the face of the CCD (charged coupled device) chip to be adjusted pixel by pixel to provide the best picture possible rather than, as in analog video processing, the application of an average value for brightness, contrast, or color.

Although signal processing within the camera is digital, the most common transmission medium is analog via coaxial cable. When digital outputs are provided, the signal is often transported by some type of transmission media (e.g., optical fiber or twisted pair).

IP Video Cameras

This is a device that produces a video image and encodes it for streaming over a network. This device combines a lens, imager, DSP (digital signal processor), and digital to analog converter in a single package. The network video camera will, at a minimum, include a connection for the network.

If this is a remotely positionable camera with pan/tilt motion and zoom lens capability, then the camera's position is controlled via commands sent from a user's computer or network recording/control command center directly to the camera over the network. In addition, the network PTZ camera may have an input/output and/or serial connector for an additional means of positioning control.

If the powered device (PD), like an IP camera, is 802.3af compliant, the power is sent over the Ethernet cable when power is requested. If it is not PoE compliant, the low-voltage (usually 12 V DC or 24 V AC or dual voltage) connections are used for power.

Network Video PTZ Cameras

Security practitioners and designers must routinely specify outdoor camera solutions for varying applications and do not have the manpower or time to test capability, resolution, performance, cost-to-performance ratio, network security, and compliance with their own company network. In addition, physical security departments are continually asked to justify expenditures by making the investment useful to other departments within the organization.

A useful solution is to select an advanced IP PTZ camera with the highest level of compliance with network standards. In addition, use of HDTV surveillance with its standardized resolution, frame rate, and color fidelity make certain cameras the most advanced outdoor PTZ devices on the market today. The security manager can shop and purchase in confidence with these capable tools.

One best practice is to verify that the camera's platform is able to support embedded applications. These platforms create a "future-proof" investment and permit video analytic vendors and solution providers to develop people-counting, object-tracking, and recognition applications and abnormal behavior recognition programs right in the camera without any outside equipment. Also, having recording capability right in the camera in the form of an SD/SDHC memory card slot is very useful in the event of network outage or to create a self-contained recording system.

A sample list of features to verify with a PTZ network or IP camera will typically include:

- High-resolution imager for day and night use
- Wide focal length range for maximum magnification
- Best illumination for both color and B/W
- Wide zoom range allows for maximum placement flexibility
- Conforms to HDTV standard in PTZ camera package
- Verify compliance with network security standards—ease of IT Department and ADJ requirements
- Verify compliance with network protocols (true IT/PhySec convergence device)
- "Green" device gets powered right over Ethernet cable—no need for external power supplies
- Internal memory recording for covert applications and in case of outage
- Withstands widest temperature ranges

As a physical security director or designer, you will try to specify the "ideal" camera; in the past, you would have wound up with several cameras, each satisfying a partial number of features. Today, you will find all these features, together with the highest level of IT and industry interoperability compliance, if you perform due diligence and make an objective selection.

Some IP video cameras include advanced features like object detection algorithms and can capture metadata at the video source for later processing by network video recorders or image processing servers. A wide range of developers offer algorithms for use in these cameras; however, there are emerging standards for interoperability framework (SIA OSIPS) and binding (ONVIF).

IP Video Encoders

These devices are often referred to as single- or multichannel video encoders or video servers. This device simply converts video from analog cameras

Figure 9.3 A network video recorder.

into multiple video streams that may be accepted by the network video recorder (Figure 9.3) or control command center. It is usually a four- or six-channel device that may be placed near the analog camera or at some distance to accommodate placement in a telecommunications room.

There are even single-channel encoders available and useful for those small analog video solutions that you wish to upgrade to IP video. The packaging may be as a stand-alone module, rack-mounted device, or card type module that is a "blade" inside a larger rack-mounted storage device.

These encoders support one or a number of types of PTZ camera control protocols. The encoder may support PTZ signals that are multiplexed over the coaxial cable or connected via serial.

Compression Technology Overview

As we have just discussed the two most popular types of IP video sources, it is appropriate that we outline the compression technologies that are used. If video compression were not used, IP video cameras and encoders would have been far less popular than they are today, because the infrastructure to support the transport of these video streams would not have been ready until recently.

IP video deployments are commonplace these days and are generally a requirement for medium- and large-scale video surveillance solutions. It is only the small video surveillance systems that have been primarily analog video deployments, but this is rapidly changing as the positive impact of hosted video solutions is making IP video deployment universally economical.

IP video solution drivers are the result of the efficient transport technologies, power over Ethernet, and the increasing efficiency of today's IP camera compression schemes. Video compression in today's IP video cameras is performed using either h.264 or MJPEG Codecs. MJPEG (motion JPEG) had been a commonly used compression technology for most IP cameras and is still commonplace to see built right into the same camera that has the capability of streaming h.264 video. With MJPEG, the camera captures still images and compresses them into a JPEG format. A camera can capture many images in 1 second (say, 30), which is the same as 30 frames per second. These images can then be transmitted over the network and viewed in sequence so that they appear just like a movie.

The latest iteration of MPEG-4, MPEG-4 Part 10, h.264, or the advanced video codec (AVC) is the most efficient compression technology to date. To view compressed video, it is necessary to decode it. This is so commonplace that mobile devices today typically decode h.264 video streams, previously thought to be too complex and processor intensive. In fact, h.264 decoding is so common that you have a nice choice of using software or hardware decoders, allowing you to view IP video surveillance virtually everywhere there is Internet or network connectivity.

License Plate Capture (LPC) Cameras

These are specialized cameras that are specifically designed to capture license plate information for processing by a license plate recognition (LPR) system. They have similar components to the IR surveillance camera package described in this chapter. Vehicle tags have a reflective coating that is sensitive to IR illumination and therefore represents an excellent opportunity to perform access control and visitor tracking in a passive manner (Figure 9.4).

Cameras with "True" Day/Night Capability

Also known as color/black and white cameras, these cameras have the capability of viewing near-IR light, usually in the range of 850 ~ 880 nm. These cameras may be used alone or together with an external IR illuminator that is sensitive in that range. Incandescent-based IR lights present danger due to the heat and infrared radiation especially when mounted at a low height. LED-based illuminators do not have this problem.

One of the most versatile features of many of today's fixed and PTZ cameras is the day/night feature. The camera that has auto and manual color-to-black and white capability first senses a low light condition that would be better accommodated by a black and white mode. The camera then

Figure 9.4 License plate capture (LPC) cameras.

automatically removes its built-in infrared cut filter, improving its sensitivity to light in that spectrum. The camera can then automatically turn on an IR illuminator and literally see in near darkness.

Cameras with Non-IR Sensitive Day/Night Capability or "Chroma Mode Capability"

Some cameras offer an increased sensitivity mode and black and white capability but do not have a removable IR cut filer. These cameras are not sensitive to IR illumination and are generally a poor selection for an IP video surveillance device.

Thermal Cameras

These cameras capture heat or temperature values of a scene and not light values, regardless of how bright or dark the scene appears to the human eye. Although the identification of colors and details is impossible with thermal cameras (because they view temperature only), these cameras are especially useful in viewing dark scenes for activities that have heat signatures.

Security practitioners and designers must routinely specify outdoor camera solutions for varying applications and need a camera that can supplement visual and physical perimeter facility patrols. The thermal imaging cameras have been too costly a solution in the past, but now products using uncooled thermal imagers might have a shorter (1,500 foot) range, but they have a price point of about 25% of the cost of cooled imager counterparts.

This advanced IP-based outdoor thermal camera can include network capability, power over Ethernet, streaming of different colored palettes to detect intruders in a wide variety of situations, and h.264 compression. Also, there are product opportunities for the designer to have thermal imaging in a platform that can permit video analytic vendors and solution providers to develop people counting, object tracking, and recognition applications and abnormal behavior recognition programs right in the camera without any outside equipment. If someone tries to tamper with the camera itself, either by redirecting it or obscuring it, an alert is transmitted or just locally recorded, along with a prerecording of who did it. Like the advanced IP PTZ camera, there can be recording capability right in the camera in the form of an SD/SDHC memory card slot.

As a guideline, here is a typical feature set for specifying an advanced, yet cost-effective IP thermal camera:

- Streaming multiple, independent, differently colored streams so that you can have the best chance at seeing an "invisible" intruder
- Video motion detection, active tampering alarm, audio detection
- Support for video analytic algorithm platforms that enable embedded applications like trip-wire or even object left behind
- Verify compliance with network security standards—ease of IT Department and ADJ requirements
- Verify compliance with network protocols (true IT/PhySec convergence device)
- Green device powered right over Ethernet cable—no need for external power supplies
- Internal memory recording for covert applications and in case of outage
- Withstands widest temperature ranges in the Americas

The ideal thermal imaging camera becomes a full-spectrum perimeter surveillance device ideal for high-risk applications like borders and embassies and is often paired with an HDTV day/night camera.

Progressive Scan CCD

This is the most powerful application-oriented imager used and is present on many IP video cameras. The progressive scan CCD is equipped with twice as many vertical transfer cells as interline CCDs, which means one vertical transfer cell for each photodiode pixel. All pixels are transferred in 1/60 of a second, resulting in full vertical resolution with less motion blur.

Lenses

As a photographer, you learn to spend as much time as possible on an arsenal of lenses, because you know you can gain big advantages in having good lenses gather the light and then project the image. In the video surveillance world, not much is different; the lens is one component that can make a big difference in improving the overall scene image quality and sensitivity.

Varifocal Lens

This lens is convenient for installation because the focal length is variable. A 2× varifocal lens can vary from wide to standard focal length and an 8× varifocal lens can vary from wide to telephoto. Unlike a zoom lens, focus needs to be readjusted when changing the focal length. This lens cannot be controlled remotely.

Lighting

The lighting requirements for a video surveillance system depend on the sensitivity of the imaging devices, the functional use of the video surveillance system, and the existing lighting at the facility. Luminance is used to describe reflected light from flat surfaces and is measured in lux (lux = lumens/square meters) and is the method of measuring the sensitivity of a video surveillance imaging device.

Since a video surveillance device depends on reflected light, those surfaces that have greater reflective capability will allow this device to better reproduce images. When designing a video surveillance system for a parking area, for example, the relative reflective properties of asphalt and concrete must be considered, with concrete being the brighter surface of the two. What does this have to do with lighting and the video surveillance camera? You will need more light with surfaces like asphalt that have lower reflective qualities to produce the same images that another camera produces near concrete surfaces.

Color Temperature

Color temperature is an important consideration when selecting lighting. Color temperature is expressed in units called kelvins (K). The color that a black body radiator glows when it is at a specific temperature is the color temperature of a light source. Some light sources are ideal for reproducing details, especially those greater than 3000 K. There are advantages to light sources that have color temperatures less than this color temperature, but they are not associated with the improved details that higher color temperature light sources produce.

Infrared Illumination

Infrared illumination is an excellent tool to add directional and flood lighting to key areas. Most day/night cameras in their black and white mode are quite sensitive to the near visible (850 nm) and invisible (950 nm) IR light that today's IR illuminators provide, making IR a low-cost way of supplying additional light at night. As stated earlier, the IR illuminator must be matched to the imaging device to assure compatibility. That said, there are some cameras that are sensitive to both wavelengths.

Pixels, Imager Sizes, and Sensitivity

As the quality of today's megapixel imagers increases, the ability to reproduce high-quality images also increases.

HDTV

HDTV is a wonderful standard that has positively impacted the video surveillance industry and therefore guarantees the expectation of video quality, frame rate, and color fidelity. HDTV is no more complex than multimegapixel imaging and in fact provides a more convenient and economical solution because video surveillance for observation, forensic review, and recognition has a higher probability for successful deployment.

High-definition imaging is an interesting topic because we use HD every day of our lives—in entertainment, photography, and, of course, physical security. HDTV is a type of multimegapixel imaging—not the other way around.

The pure definition of HDTV (high-definition television, or HDTV, or just HD) refers to video having resolution substantially higher than traditional television systems (standard-definition TV, or SDTV, or SD). HD has one or two million pixels per frame—roughly five times that of SD.

HDTV provides up to five times higher resolution than standard analog TV. HDTV has better color fidelity and a 16:9 format. The two most important HDTV standards today are SMPTE 296M and SMPTE 274M, which are defined by the Society of Motion Picture and Television Engineers (SMPTE).

HDTV broadcast systems are identified with three major parameters:

- Frame size in pixels is defined as number of horizontal pixels × number of vertical pixels—for example, 1280 × 720 or 1920 × 1080. Often the number of horizontal pixels is implied from context and is omitted.
- The scanning system is identified with the letters "P" for progressive scanning or "I" for interlaced scanning.
- Frame rate is identified as number of video frames per second. For interlaced systems, an alternative form of specifying number of fields per second is often used.

If all three parameters are used, they are specified in the following form: [frame size][scanning system][frame or field rate] or [frame size]/[frame or field rate][scanning system]. Often, frame size or frame rate can be dropped if its value is implied from context. In this case the remaining numeric parameter is specified first, followed by the scanning system.

For example, 1080i30 or 1080i60 notation identifies interlaced scanning format with 30 frames (60 fields) per second, each frame being 1,920 pixels wide and 1,080 pixels high. The 720p60 notation identifies progressive scanning format with 60 frames per second, each frame being 720 pixels high.

Images and video clips from HDTV and quality multimegapixel devices are more usable when they come from devices that meet standards. Public safety professionals are now using HDTV because they have seen demonstrations and sample video clips that demonstrate use cases, like monitoring critical events for first responders, operating room surveillance, and transportation security.

The deployment of MESH wireless networks and the availability of network infrastructure of all varieties will not only encourage use of HDTV, but also make it a device that is normally deployed, right alongside a network switch.

Adoption of HDTV is taking place for nonsecurity applications even more rapidly than for security surveillance; it just depends on the industry you work with. Markets with a high degree of penetration include:

- Medical (operating room, remote diagnosis)
- Transportation (airport surveillance)
- Entertainment and resort (advertising)
- Sports, both surveillance and player coaching

Surveillance and security require that operators need better situation awareness and HDTV's image quality delivers that. As metrics are discovered that illustrate how operators are able to perform their jobs more effectively with HDTV, this will accelerate its adoption.

Automated recognition-based systems that use video analytics perform more effectively on an HDTV platform, and these cameras typically have stronger processors, so they can run analytics and stream video effectively. Video analytics allow a guard force to be more aware of off-normal events by allowing HDTV systems to handle the burden of many routine tasks. For example, an HDTV camera, together with an access control device, can admit returning contractors without the need for operator interaction. Security operators can be proactive with other tasks, reducing manpower costs.

Some of the best use cases for HDTV network video surveillance include the following:

- City center surveillance
- Medical (operating room, remote diagnosis)
- Transportation (airport surveillance)
- Entertainment and resort (gaming and advertising)
- Sports, both surveillance and player coaching
- Education—distance learning
- Any application requiring the consistent delivery of high-quality imaging for both live observation and recorded use

There are many advantages to deploying HDTV IP video surveillance. There used to be a larger price difference between non-HDTV and HDTV cameras, but this has been greatly reduced. With the use of h.264 video compression, bandwidth consumption is also reduced. Packaging of HDTV cameras is now available in fixed box style, fixed dome, and PTZ camera form factors, so this has also been accommodated. The only area that HDTV cameras have not been able to impact is the OEM camera market, such as ATMs and automotive and marine safety, but this is expected to change in the next couple of years.

If the use case requires higher resolution or more pixels on target, then market share will increase from that application. Video analytic algorithms run better with higher quality imaging. Forensic review gets performed more easily.

Public safety, law enforcement, and medical applications all benefit from the use of HDTV video, and with more efficient compression technologies like h.264, the deployment of HDTV is often a first choice when selecting an IP camera.

Different IP cameras, HDTV or not, use different compression engines and can therefore at times be an interoperability challenge—but one that is overcome through the use of a standardized API. In addition, interoperability alliances like ONVIF create common ways for IP cameras to bind with the video management system. The structure of communication between video surveillance devices has been standardized by the Security Industry Association in its ANSI-approved OSIPS standard.

HDCCTV

High-definition closed-circuit television is built on technology pioneered for broadcast television. Previously referred to as "analog megapixel," video is transmitted uncompressed and without being encapsulated in TCP/IP. The result is a system in which a camera can be plugged into a receiving device and video can be displayed without latency and minimal configuration. HDCCTV recording and display resolution is dependent on a number of factors that do not adversely affect HDTV's image quality. HDCCTV makes promises to bring all of the benefits claimed by megapixel IP cameras to the video surveillance market with conventional analog video surveillance equipment.

HDCCTV is a point-to-point system and claims not to require any additional infrastructure to deploy. This is not true if you want full frame rate and especially if you have older and poor quality cable.

New and existing installations claim to be able to use video surveillance industry standard coaxial cable (RG/59, RG/6 and RG/11). However, it has yet to be proven whether existing coaxial infrastructure may be used to transport the signals from non-IP megapixel cameras adequately.

At best and assuming that you have perfect cable, pure copper center conductor, pure copper braid overall shield with 95% or greater coverage, and three-piece professionally crimped BNC connectors, here are the length limitations: If you have perfect RG59U cable, HDCCTV requires continuous cable runs less than 300 feet. If you have perfect RG6U cable, HDCCTV gets you a little further to 800 feet.

There are Ethernet extension units and Ethernet over coaxial media devices (example: Veracity's Hiwire and Outreach devices, respectively) that permit runs of thousands of feet, over ordinary cable and far less concerned about cable quality.

HDCCTV claims to be designed to be a drop-in replacement for existing analog video surveillance, requiring only a change of camera and receiver; however, enhanced DVR capture cards may be required to accommodate non-IP megapixel cameras. HDCCTV is designed to be forward- and backward compatible, meaning that early adopters will be able to continue to use HDCCTV equipment as more features are added to the specification.

HDCCTV operating at 720P provides almost three times the video resolution of analog video surveillance, and 1080P provides six times the resolution, but it is dependent on infrastructure and analog capture cards to deliver the expectation of this resolution.

Yes, all HDCCTV systems are progressive, eliminating the flicker and blurring associated with conventional analog video surveillance systems. However, there are significant and hidden costs and possible field qualifiers that need to happen prior to deployment.

The process of adding one HDCCTV camera to a video surveillance installation can be more complex than adding one megapixel network camera to a video surveillance installation. Deployment of HDTV or IP megapixel cameras is indeed simple with the use of Ethernet cables and a network switch. The use of HDCCTV cameras may require more specialized DVR and video capture cards to accommodate the video sources' increased resolution, not to mention the variable nature of coaxial infrastructure.

The biggest issue with HDCCTV is adoption by manufacturers. IP-based HDTV solutions are now adopted and manufactured by companies holding the largest market share of standards-based IP video. HDCCTV, at the time of this writing, does not have this type of adoption by major manufacturers.

HDTV Deployment Justification

Matching the video source to the surveillance function and performing a good video site survey are ways of minimizing design challenges in an HDTV video surveillance solution. How is a HDTV video surveillance a good ROI and reduced TCO (total cost of ownership)?

- There is a cost savings of the use of existing IP infrastructure
- Use of the HDTV system is very suitable for other purposes than security, like safety, operations management, and education
- There is a measurable increase of useable video when HDTV devices are deployed. For example, if we even consider the split of observation/forensic review/recognition as 40%/40%/20% of a customer's applications, then a standards-based 720p HDTV system will deliver images that may be four times the quality and improve 60% of a customer's images

Recording Systems

Digital Video Recorders

To the consumer, anything video- or image captured digitally and stored may be classified as a digital video recorder, or DVR. To the security industry, a DVR is a multichannel analog input device that captures composite video, stores it on digital media, and permits search, playback, and video analysis functions for single or multiple users. Multiple capture cards or DVR cards, together with a server and digital video management software that also does the above, could collectively be a DVR.

DVRs are a multichannel device with a proprietary, embedded, or built-in real-time operating system that acquires video images at a preprogrammed format and capture rate and writes and stores the same data in a proprietary format on nonremovable hard disk drives.

Network Video Recorders

Network video recorders (NVRs) represent an important, growing product segment for the deployment of video over both dedicated and shared networks. We see versions of these recorders in our everyday lives; home digital video recorders provide convenient ways of unifying the source selection and schedule recording process.

Video Management Systems

These systems are usually multiuser, have software costs based on camera licensing, manage video streams and store video on dedicated or shared servers, and permit and manage multiple users to view live and review prerecorded video. The video management system (VMS) manages video as data and therefore performs searches based on various time, location, and related event criteria. The VMS software can be locally delivered or virtually served from "the cloud" and usually manages the video data enabling search requests that can span across disparate video data storage locations.

Servers running VMS software must be sized to accommodate the number of video streams, number of concurrent users, processing power needed to perform concurrent search requests, alarm input activations, and foreign system controls.

Video Surveillance and Cloud Computing

In order to understand video surveillance's evolutionary move to hosted environments for particular use cases, we need to define cloud computing and virtualization first. Cloud computing is a model for enabling ubiquitous, convenient, on-demand network access to a shared pool of configurable computing resources (e.g., networks, servers, storage, applications, and services) that can be rapidly provisioned and released with minimal management effort or service provider interaction. (This definition is from the latest draft of the NIST Working Definition of Cloud Computing published by the US government's National Institute of Standards and Technology.)

There are three major ways to categorize delivery methods of the cloud's software and infrastructure services: private, enterprise, or public cloud delivery. So now that we understand delivery, let us tackle the three most popular types of service: software as a service, infrastructure as a service, and platform as a service.

Software as a service (SaaS) has most positively impacted video surveillance by moving the software to a hosted or managed video portal. These services are delivered by video hosting service providers and deliver some signification advantages to certain video surveillance applications.

Here is a common problem: There is a massive aging population of analog digital video recorders that are either partly or nonoperational. Should the security practitioners and designers replace these systems with newer versions or can they save money by leveraging cloud computing security solutions designed for multiple locations of small- and midsize installations?

The Solution

The video hosting systems can deliver monitoring and recording via cloud computing (software as a service) and still use the latest camera technology to the end user via hosting providers at a substantial cost savings. Some of the advantages that end users will experience for these geographically dispersed small systems include:

- Automatic binding of camera to hosting site allows for a simpler installation.
- All video is stored in the cloud, so there is no evidence to lose.
- High-definition video is recorded directly from camera to an optional network-attached storage device, placed anywhere at the facility.

The three steps to binding the hosted video camera to a portal are as follows:

1. Connect the camera a to network switch or WiFi router.
2. Program camera identification information at the monitoring facility or central station.
3. Monitor real-time or recorded video on any platform, including laptop, desktop computer, or mobile device, via a browser.

There are significant benefits for deploying these solutions:

- No software upgrades or antivirus software required for duration of service
- Mobile and remote devices like Blackberry, Android, iPhones, iPads, and laptops are supported directly from Internet site; you still get alarms, real-time and recorded video even if local storage is damaged

- Reduced installation time
- Works with existing infrastructure four ways:
  - Existing analog cameras
  - Wireless Ethernet
  - Wired Ethernet
  - Ethernet over power line

When compared with a replacement DVR system, the user can save over 50% for a 4-year period, or about $5,000 (currently) for a four-camera system. HDTV cameras have a higher bandwidth stream that is sent locally to a network attached storage (NAS) device. The NAS is also there to record if the Internet connection ever goes out. Larger end users can deploy this on their own networks and achieve even greater savings.

Hosted video solutions represent a growing market as they are the lowest cost, most technologically superior, and easiest solution for video surveillance systems that are small in size and geographically dispersed (see Figure 9.5). The user simply needs to decide whether to deploy the solution himself or herself or through service (hosting) providers. Aging DVRs that require an upgrade can be complete in 25% of the time, compared with conventional analog solutions.

Whether infrastructure as a service (IaaS) is storage or computing power, companies providing these cloud solutions are a great complement to SaaS. The IaaS customer is a software owner that is in need of a hosting environment to run software. IaaS vendors use virtualization technologies to

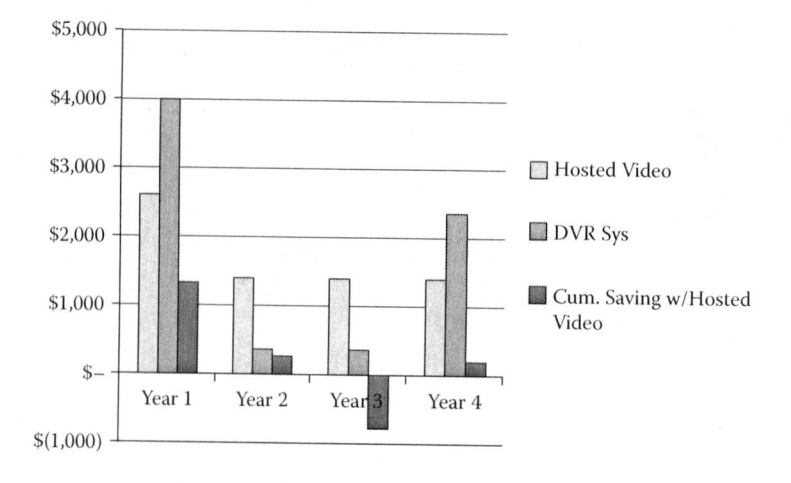

Figure 9.5 Typical hosted video savings. Graphic is based upon a four-camera hosted video system including a remote device using a DVR system with HDD maintenance.

provide computing power. The unit of deployment is a virtual machine built by the software owner. A virtual machine is built for an IaaS environment, uploaded, configured, and then deployed within the environment. A video hosting provider can use IaaS as cloud storage to scale the solution. IaaS providers represent excellent opportunities because of their experience in providing e-services, a key method of delivery.

Platform as a service (PaaS) is a whole new breed of companies offering a dev platform for hosted solutions. A customer of a platform as a service offering is also a software owner that is in need of a hosting environment for his or her application. PaaS provides computing power by providing a runtime environment for application code.

Public Cloud

In simple terms, public cloud services are characterized as being available to clients from a third-party service provider via the Internet. The term "public" does not always mean free, even though it can be free or fairly inexpensive to use. A public cloud does not mean that a user's data are publicly visible; public cloud vendors typically provide an access control mechanism for their users. Public clouds provide an elastic, cost-effective means to deploy solutions.

Private Cloud

A private cloud offers many of the benefits of a public cloud computing environment, such as being elastic and service based. The difference between a private cloud and a public cloud is that in a private cloud-based service, data and processes are managed within the organization without the restrictions of network bandwidth, security exposures, and legal requirements that using public cloud services might entail. In addition, private cloud services offer the provider and the user greater control of the cloud infrastructure, improving security and resiliency because user access and the networks used are restricted and designated.

Community Cloud

A community cloud is controlled and used by a group of organizations that have shared interests, such as specific security requirements or a common mission. The members of the community share access to the data and applications in the cloud.

Hybrid Cloud

A hybrid cloud is a combination of a public and private cloud that interoperates. In this model, users typically outsource nonbusiness-critical information and processing to the public cloud, while keeping business-critical services and data in their control.

Service providers or video hosting providers operate a server platform providing services directly to end users or through service resellers or video service providers. Sometimes this terminology is confusing since the IT industry uses the term "service provider" for what the security industry calls a "video hosting provider."

Video service providers qualify the application, match it with a use case, and sell the hosted solution. They typically offer the hosted video providers' services to end users and maintain the system at the remote site.

Video Control, Analysis, and Video Content Analysis Systems

Video Analytics and Automated Object Detection

Video analytic systems can include object recognition systems like facial, LPR, object left behind, and people counting. They are also used to recognize behavior without determining object recognition, as in trip wire perimeter detection, auto-tracking, temporal motion analysis, behavior, and speed.

Users are finding that video analytics require qualification by trained systems integrators and the initial deployment may be different from accuracy claimed on specs—hence, an often poor result of false positives, false negatives, and eventual ignorance of the alarm conditions. This is why a strong, accurate, qualified video analytics engine, combined with dark screen monitoring, can be your very best video surveillance solution at the command center.

Manufacturers of camera platforms, analytic subsystems, and the VMS need to be intimately involved in the deployment process to provide intensive training. Here is a good guideline/cycle:

- Application and systems integration qualification
- Deployment method
- Correction
- Redeployment

There are applications that use video analytics and can perform complex repetitive functions like object detection and recognition on many channels of video simultaneously. The designer should consider a system like this in the following circumstances:

- The system uses a large quantity of cameras that require monitoring for specific conditions or behaviors that are capable of being recognized.
- The setup and installation of the video analytics subsystem is relatively simple and has high, sustained accuracy for the types of behaviors and objects recognized.
- Modifications may be made by the end user easily and effectively without the constant involvement of the integration professional.

When selecting video analytics and object detection subsystem, the video surveillance designer should consider the use of processing a metadata feature stream at the camera and then sending it to a server that analyzes and has a user interface supporting criteria search. This search result is typically a series of thumbnail images of video clips that reside on the video management server. These solutions are now very useful if the designer is able to select a camera manufacturer that has a ubiquitous platform to house the onboard algorithm and enough processing power to run this and still stream high-resolution (HDTV) video.

Content Analysis

Video content analysis is the action of searching through video data and returning a result, based on criteria. There are two major types of VCA: intelligent video search and video synopsis. There is a massive amount of video that the security practitioner has to review if an incident occurs. This review takes time and keeps manpower from doing the main job: protecting facilities from threats.

Intelligent Video Search

An open camera analytics application platform allows intelligent algorithms to gather data. An intelligent video search server-based application enables effective analysis, retrieval, and presentation of specific video segments, events, and data from vast amounts of recorded video.

A typical intelligent video search application might

- Automate search and analysis processes to enable leveraging of stored video
- Scan days of stored video in seconds to display precise results
- Distribute automated alerts and countermeasures if a critical incident happens
- Offer high search flexibility, including searches for behavior type, target size, and color
- Search simultaneously on stored video from multiple cameras
- Work alone or with the existing video management system
- Significantly increase ROI of your video surveillance infrastructure by making it useful to other departments
- Search for entities through application of advanced search parameters, like vehicle color, type, speed
- Provide a graphical presentation of all motion paths in a scene, with access to related video segments stored at the VMS server
- Conduct people counting or vehicle analysis; generate statistical reports

Consider that lives in a vehicle traveling in the wrong direction down a thoroughfare could be saved by a solution that combines intelligence at the video edge (camera) combined with appropriate real-time visual interdiction. An investigation requiring the review of days of recorded video will be reduced so significantly that law enforcement can react in near real time. A solution with intelligent video search might make this possible.

Video Synopsis

The typical video archive search can take time and have definable costs with the loss of video information or response delay to an event. Should this error or oversight be tied to a significant attack or exploit, the damage can be devastating. Video synopsis or the summarization of video can present a simplified event-based clip accurately showing an entire day's motion or other criteria-based events in several minutes. Once an event is identified through the synopsis, indexing to the full video content can occur as quickly as access to the VMS server is permitted.

The wonderful thing about this type of application is that it is elastic and may be applied to a single IP camera or video stream or to many. Do you want to record the construction of a building over a long period of time? Synopsis is a new, more effective way to perform this task than time-lapse imaging.

Synopsis can also display just the objects or behavior that you wish to monitor. Sample video synopses might include

- Display only vehicles moving from left to right (and vice versa)
- From the same video footage, display pedestrians only
- Display summarized video footage from a tracked object or class of objects

Synopsis can be a most useful tool for facility security. A guard force manager might review synopsis at every change of shifts. In a shopping mall, a stolen purse leads directly to an examination of the last 30 minutes of video inside the mall and outdoors in the parking area, simultaneously, in under 1 minute. Systems that are designed primarily for forensic review benefit directly from this technology by reducing the review time after an incident or during a routine shift review.

The most significant use of this technology permits law enforcement to review video content leading up to suspicious behavior quickly to possibly avert disaster.

Interoperability

Interoperability is concerned with the ability of systems or video devices to communicate. It requires that the communicated information be understood by the receiving system. In the world of the IP camera, this means providing a standardized API to the VMS or recording system manufacturer and accommodating basic features like video stream decoding, video motion detection alarm processing, PTZ camera controls, input/output processing, analytics module feature stream decoding, auto discovery and binding, automatic IP address assignment, bandwidth management, and general detection of the presence of the IP-based device.

The IP camera manufacturer must achieve 100% compliance with standardized network protocols and deliver the API for a given camera in advance of its product release so that the VMS manufacturer can provide a total solution.

In the world of cloud computing, this means the ability to write code that works with more than one cloud provider simultaneously, regardless of the differences between the providers.

Integration

Integration is the process of combining multiple components or systems into an overall working solution. Some examples can include an access control system requesting video clips from door held open events, an intrusion detection system recalling video clips leading up to a late-to-close burglar alarm event. In addition, solutions like physical security information management systems (PSIM) can be the integration gateway for disparate systems and provide a common and comprehensive user interface.

Integration among cloud-based components and systems can be complicated by issues such as multitenancy, federation, and government regulations.

Application Programming Interface

An application programming interface (API) is a contract that tells a developer how to write code to interact with some kind of system. The API describes the syntax of the operations supported by the system. For each operation, the API specifies the information that should be sent to the system, the information that the system will send back, and any error conditions that might occur.

Using a Step-by-Step Approach to System Selection and Deployment

The video surveillance designer is encouraged to use the following framework for designing an analog or IP video system. Each of the individual tasks is presented as an overview and may require additional detail:

1. Plan the solution
   a. Perform an assessment process to determine the use of the video surveillance system

b. Identify and rank critical assets
c. Identify vulnerabilities and develop responses
d. Assessment
e. Analyze network architecture
f. Assess threat environment
g. Examine policies and procedures
h. Assess Infrastructure Interdependencies
i. Identify risks
j. Are there any physical vulnerabilities to your infrastructure?
k. Have you protected against system attacks from within your facility?
l. Capture lessons learned and best practices
m. Conduct training
n. Classify video system by type:
    i. Observation
    ii. Investigation and forensic review
    iii. Recognition of objects, such as human faces and license plates
    iv. Prosecution
    v. Loss prevention and deterrence
    vi. Intrusion detection and perimeter monitoring
    vii. Access control
    viii. Operations management and resource allocation
    ix. Safety
    x. Security/video surveillance (monitoring function)
o. Conduct your best site survey
    i. Assess lighting conditions
    ii. Associate a function to a camera
        A. Monitoring
        B. Detection
        C. Identification
        D. Recognition
        E. Access control
    iii. Note environmental conditions and possible vandalism
        A. Viewing angle
        B. Environment
        C. Lighting type, level
        D. Reflectance/emissivity
        E. Obstructions
        F. Temperature
        G. Humidity

H. Corrosives
I. Water
J. Vandalism
   iv. Note "special" objects/scenes and match imaging devices and video management solution, deploying video analytics or video content analysis
      A. Vehicles (use HDTV devices)
      B. License plates (use LPR and LPC application with fixed camera or mobile LPR system if attached to patrol vehicle)
      C. Shoreline/border (consider using thermal imaging)
      D. Transit platforms (use vandal-proof HDTV interfaced to emergency call stations)
p. Video site survey cycle
   i. Enter site survey data (rough sketches, location notes)
   ii. Use a standard site survey sheet
   iii. Normalize devices by camera types
   iv. Transfer to floor plans
   v. Verify samples with actual site conditions (see following task)
q. Verify all monitoring and recording functions using resolution targets (see broadcast television test charts and digital imaging reference charts for best examples)
   i. Adjust system as necessary for key areas, adding or moving cameras as required
   ii. Sample equipment for video site survey and test
      A. Laptop computer
      B. Fixed "box" type IP camera
      C. Long Ethernet stranded crossover cable
      D. Assortment of varifocal lenses
      E. Tripod/ball head for camera
      F. Light meter with gray card
      G. Sample ISO12233 I3A/ISO standard test target chart
      H. LED infrared illuminator/power supply

2. Select camera types and imaging requirements
  a. Match camera function and desired resolution
  b. Assess difficult lighting conditions
     i. High-contrast lighting (wide dynamic range use)
     ii. Day/night switchover
     iii. Automatic lens focus or back focus
     iv. Image stabilization requirement

      c. Imager resolution size
          i. Match resolution size also with required bandwidth allocation
          ii. MegaPixel IP cameras can require substantially more bandwidth
      d. Consider certain technologies to aid identification
          i. Consideration in the site survey process
          ii. Face location/face detection
          iii. Auto tracking/auto zoom with PTZ cameras
          iv. Use of dynamic imaging/backlight compensation for high-contrast lighting situations
      e. Deploy PTZ cameras to match monitoring requirements, especially in larger coverage areas, primarily activated by wide-view HDTV cameras, and then for investigation of an alarm, with automated home position return
      f. Use lens estimator to size lens
          i. Enter scene height, width and distance for each camera
          ii. As an alternate, note actual viewing angle on scaled diagram
          iii. Determine object height on monitor to match functional requirement
          iv. Determine focal length
3. Determine initial system selection type
      a. Match number of cameras to recording system type
          i. An example of a small recording system is one with eight or fewer network cameras connected to a small network attached storage (NAS) device and connected to a hosted video provider when an Internet connection is present
          ii. A midsized IP video system with multiple NVRs is capable of handling higher throughput and more users managed from a single or multiple workstations
          iii. Should the need arise to connect existing or specialized analog cameras, use multi- or single-channel encoders where necessary, achieving an IP-based system; replace analog cameras in the future as required and when budget permits
      b. Choose between embedded system and server-based system
          i. Embedded systems usually have no camera licensing costs, have a fixed network throughput, and low ability to interface to foreign systems
          ii. Server-based recorders require camera and/or software licenses, have variable network throughput, require additional software to protect files and require a skill set to maintain, but have the

highest degree of interoperability, video management, and ability to support video analytics and video content analysis
    iii. Verify camera protocol compatibility and API for any embedded or server-based system
  c. Determine recorder capacity or external storage capacity required, leveraging existing storage infrastructure or private or public cloud infrastructure
  d. Match recorder type and quantity with number of users accessing the video information
  e. Also match network switch type and network throughput and PoE capacity
4. Calculate bandwidth and storage values
  a. Use multiple manufacturers' bandwidth and storage calculators, in addition to server estimators
  b. Populate use case form (see figure), matching closest application
5. Develop bandwidth/network/routing maps
  a. Develop bandwidth measurement scenarios
  b. Note individual camera bandwidth values
    i. Use camera and recorder bandwidth calculators as required
  c. Accumulate to nearest network switch; apply totals
  d. Accumulate multiple network switch bandwidth; apply totals
  e. Accumulate total camera bandwidth for each network video recorder; apply totals
  f. Perform user identification by function
  g. Note individual user monitoring station bandwidth values
  h. Accumulate to nearest network switch; apply totals
  i. Accumulate typical multiple user monitoring station bandwidth values for each network video recorder; apply totals
6. Verify infrastructure compatibility and protocol support
  a. Cable infrastructure needs to support bandwidth capacity requirement
  b. Cable installation and cable quality significantly impacts data rate
  c. Wiring plant topology and network switches need to support placement of security and surveillance system devices
  d. Deploy power over Ethernet systems effectively to complement your system's requirements
7. Apply topology requirements
8. Encourage "convergence" and use professional imaging over networks using the most qualified and professional product possible. Make the

video available to as many users as possible that have permission to view and control the video streams.
9. Make resolution adjustments as required to meet end user requirements; revise bandwidth allocation maps
10. Verify recording, control and management system design
    a. Verify recording system utilization with recording streams, monitoring users
    b. Make sure users have the live and recorded image quality they need
    c. Maintain most favorable criteria for your project
    d. Maintain lowest reasonable cost for system
    e. Increase the number of ways you can access your video information
    f. Strengthen and secure network devices
    g. Do not have one device service to many users or perform many functions; improve your system survivability
    h. Model all network-based systems prior to deployment and correct system design as required (see following items)
11. Check common network usage scenarios
    a. Recording system utilization with recording streams, monitoring users
    b. Network switch functions to support bandwidth load and failure scenarios
    c. Provide recovery from the most common and reasonable infrastructure failures
    d. Simulate as many network conditions and loads as possible for components, edge devices, and infrastructure
12. Develop a comprehensive performance specification for a video-based surveillance system
    a. Verify inclusion of construction specification institute sections
13. Publish a network video commissioning statement to specify the system deployment
    a. Step-by-step staging, programming, installation, and commissioning tasks
    b. Division of responsibilities
    c. Acceptance test criteria
14. Preproject diagram development
    a. Security block diagram*
    b. Data closet design*
    c. Camera schedule*
    d. Camera detail
    e. Wiring plant diagrams

  f. Point-to-point diagrams
  g. Command center elevations and stretch-out*
  h. Command center sequence of operations by scenario
15. Develop master bill of material
  a. Camera bill of material
  b. Telecomm room BOM
  c. Control system BOM
  d. Workstation BOM
  e. Command center BOM
16. Project diagram development
  a. Security block diagram (update as required)
  b. Data closet design (update as required)
  c. Camera schedule (update as required)
  d. Camera detail
  e. Riser diagrams
  f. Point-to-point diagrams
  g. Command center elevations and stretchout (update as required)
  h. Command center sequence of operations by scenario
17. Project management
18. Testing
19. Acceptance test based on specified expectation of video surveillance system performance (ask responsible integrator for his or her own suggestions on acceptance test criteria, then modify/append as required after end user review)
20. Framework for continuous performance verification
  a. Performance testing metric
  b. Positive influence against possible legal challenges to video evidence
  c. Stay up to date with current technology, continually measuring strength of systems use in facility management or PhySec program; talk to peers in similar capacity and similar industry
  d. Stay informed on risk change based on legal precedent, especially with premises liability cases
  e. Keep up to date with AHJ and industry certification entities; seek to grandfather solution as necessary to reduce costs

---

* Usually required for planning and design phases of video surveillance project; all other diagrams required for installation and commissioning phases.

Upgrade Path

The video surveillance designer should consider the life cycle of the solution and always keep track of technology refreshes available and their possible savings.

When and how to upgrade can have great impact on the design process. The video surveillance designer is encouraged to model his or her design in various situations that the user is planning to experience or some that the user has not even considered. Technology is continually evolving and the end user will continue to rely on the video surveillance designer in years ahead to interpret which technologies are innovative and how they may be integrated into the video surveillance system.

Here are some considerations when considering a video surveillance system upgrade:

- The path to an IP video upgrade leads to great benefits and can work over virtually any infrastructure with the appropriate media converters.
- Should the upgrade include interoperate, integrate, or interface to various systems?
- Should the user postpone interoperability? What are the ramifications?
- Should the running of concurrent systems be a consideration?
- Are quality of service and fault tolerance improved with the upgrade?
- Consider logical and network security issues for the integration upgrades.

For long-term viability, for those who are not ready to replace their entire system, new technology can and should be integrated with the systems and technology in place. There is a balance that must be maintained between user needs and reasonable costs for at least partial system upgrades that are necessities.

Both financial and operations efficiencies can be realized through legacy system upgrade and a technology refresh. Keep what works and enhance/replace with solutions and devices that contribute directly as relevant tools for the end user (see Figure 9.6). A great example of this is HDTV surveillance that permits live and recorded video to be more easily used as the limitation of image quality is virtually eliminated when the correct, upgraded product is matched for the correct use case.

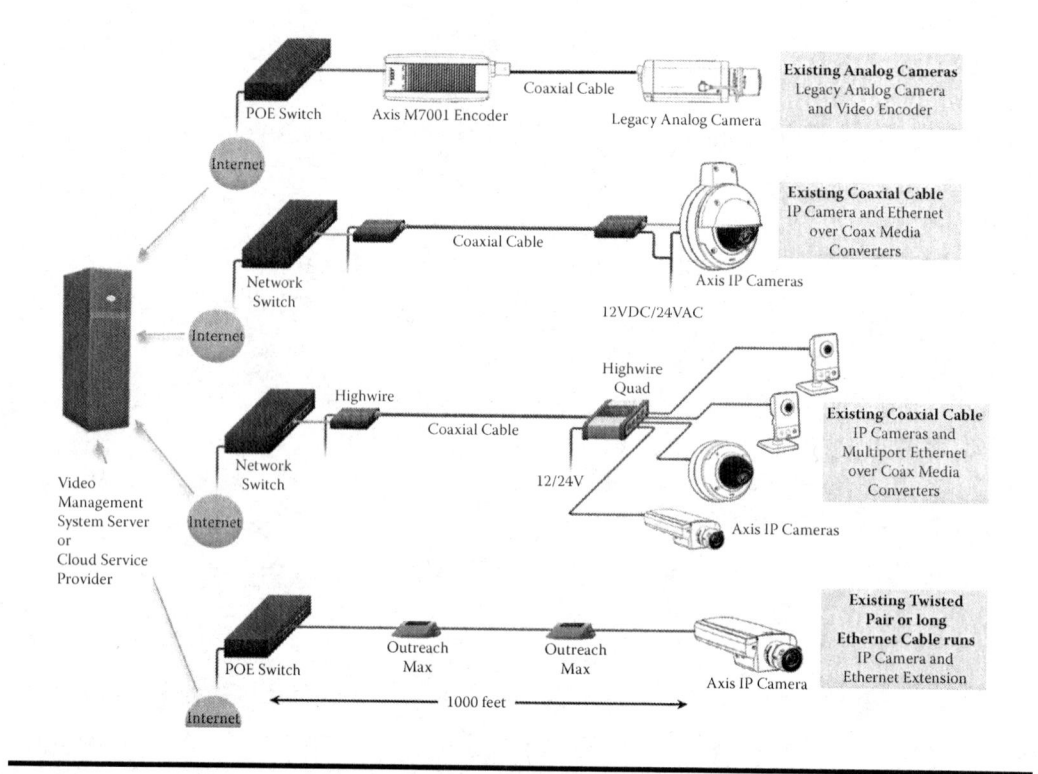

Figure 9.6 With the costs of IP and networked camera options going down, the cost savings to convert from traditional CCTV are becoming less of a barrier, even for small-scale, multicamera implementation.

Manpower and Video Surveillance

There is always a continuing, controversial discussion about whether technology should be used to replace or enhance the security guard force, and how video surveillance can specifically be used. There are multiple, global case studies showing that manned security has been significantly enhanced and, in some cases, replaced with video-based technology.

Will the balance of manpower and technology change in the next 10 years? The answer to this question is hopeful in that manpower will most likely not be replaced, but displaced and integrated into the culture of a given company. As technology grows into a more integrated tool for the security officer, guard force reduction will climb until security guards become more of company ambassadors who are trained and integrated into company culture. When this occurs, security guards will no longer exist as a single role, but will be a duty or part of a job description of a class of employees. It is a logical evolution of efficiency as these multifunctional employees learn to use technology to better understand physical security and interface with law enforcement.

There is evidence, however, of technology challenging guard force manpower. In 2007, a New York City court case ruled in favor of video surveillance to replace security guards:

> A landlord asked the DHCR for permission to modify building services. Landlord sought to replace a parking lot security guard with a state-of-the-art video surveillance system throughout the entire development. The DRA ruled for landlord. Tenant appealed, claiming that video cameras would not prevent security-related incidents. The DHCR ruled against tenant. Tenant submitted no proof that the installation of the security cameras and monitoring station would reduce security services.[1]

In a Carter/ADT case study,[2] guard force reduction was associated with a system deployment:

> But is it necessary for a security guard force provider to be able to offer electronic security solutions as part of a total security offering? It is an essential tool as the provider needs the technology to allocate manpower and perform surveillance functions more efficiently. In this case it is preferable for a manned security provider to partner with established electronic security companies. The guard force company gets to work with standardized products that can be most easily supported by the provider and end user working as a team. Also, the standardized COTS products will allow for easier upgrades and technology refreshes, keeping pace with the changing PhySec industry as it converges more with logical security.

Can technology help a security guard company solve its customers' most pressing issues? Yes, when the manned security force is integrated into a company's culture, becomes multifunctional, and is better trained to operate and use high-technology tools, it will be empowered rather than replaced.

Training for end users and their security manpower is a continuing issue. How should this task be divided between the equipment/software suppliers and the security integrator and what could be done to radically improve training and technical support for end users? In many cases, training is already available directly for end users. Sales and service channels for the delivery of security devices and security maintenance and service vary by company, market type, geographic location, and the stage of evolution that a company is in with regard to recognizing that each employee's function has a direct impact on the product or service that is provided to an end user. Since the end user population far outnumbers the security integrator's and manufacturer's resources, education methods must be standardized, driven by end user needs, be interesting, lead to a desired credential and personal advancement, and ultimately be part of a learning management system (LMS).

Privacy and Legal Considerations— Video Surveillance Checklist

Video surveillance continues to be a sensitive issue in many workplaces. Employees may challenge video surveillance on a wide variety of grounds—some of them based on federal or state laws and some based on court decisions. Because the law in this area can vary considerably from state to state, the most important step an employer can take to avoid challenges to video surveillance is to consult a knowledgeable attorney before undertaking such surveillance.

The following 14-item checklist[3] is a good starting point and includes other suggestions that may help employers avoid (or win) legal challenges to video surveillance:

1. Use video surveillance only when justified by a legitimate business purpose (e.g., preventing theft or workplace violence, investigating illegal or improper conduct, monitoring employee performance).
2. Limit video surveillance to the least intrusive time, place, and method that will serve the business purpose.
3. Use only visible cameras or inform employees in writing that hidden cameras may be used.
4. Obtain written employee consent to video surveillance for legitimate business purposes.
5. Do not use video surveillance in areas where employees have a reasonable expectation of privacy (e.g., restrooms, locker rooms, dressing rooms, lounges, employees' homes, or other places outside the workplace where employees are not in public view).

6. Do not use video surveillance devices that capture or record sound without complying with federal and state wiretap laws.
7. If the workplace is unionized, comply with any provisions in the collective bargaining agreement concerning video surveillance; if there are no such provisions, negotiate with the union before implementing video surveillance.
8. Do not use video surveillance in connection with union activities (or other "concerted activity" by employees concerning terms or conditions of employment).
9. Do not select employees for video surveillance in a manner that might be considered discriminatory under federal or state discrimination laws (e.g., do not videotape only women or only Muslims or only people with disabilities).
10. Do not select employees for video surveillance in retaliation for exercising rights under any law.
11. Determine whether the company is subject to a state or local law dealing with video surveillance and, if so, comply with that law.
12. Train supervisors in the legal issues involved in video surveillance.
13. Treat information obtained through video surveillance as confidential, and limit access to video recordings to security personnel or management personnel with a "need to know."
14. Adopt procedural safeguards to avoid unintended or improper use of work-related video recordings.

The Future of Video Surveillance

It is always fun to imagine what will be useful to users of video surveillance systems in the future. It is important to realize the great effect that key technologies adopted by the consumer electronics industry have on other professions, including physical security. HDTV is one example of a technology having its source in professional video, being widely adopted by consumer electronics, and now positively impacting the PhySec industry. Power over Ethernet, from the telecommunications and data transport industries has given us a way to power indoor and outdoor cameras with a single Ethernet cable.

One possibility might be in the way in which cameras communicate with each other. Cameras that become intelligent by way of small programs or specialized algorithms written for them will, when multiplied, become an intelligent "security net" or "intelligent hive" that can communicate in a peer-to-peer manner—analyzing and tracking potential threats, dynamically reallocating collective computing power, and learning from the results. The collective intelligence of years of metadata and event results will work for the end user and be proactive in providing the most appropriate response to a given situation.

Also, today's individual video-based object recognition, which can retrieve synopses of video clips that correspond to criteria, will automate the reaction process by presenting video footage before and after events, thus making the video management system imitate repetitive review tasks and therefore reduce manpower needs. Manpower will be better allocated as a result of this; for example, statistical temporal correlation of domestic violence cases linked to times of the month or year will "sensitize" public surveillance tools in operation and will move the "eyes" of law enforcement to areas needing it. If domestic violent crimes are associated statistically with temporal and location data, patrol vehicles will have suggested routes to move through and act as "force multipliers" in high-risk areas.

What could be the impact on manned security? The direct impact will be to move the job of security to a smarter, better armed and allocated law enforcement agency that cooperates with the public and shares the surveillance function.

References

1. Location: 2940 Ocean Parkway; source: DHCR Adm. Rev. Docket No. TE210064RT (1/18/07).
2. http://www.adt.com/medium_large_business/reference_library?wgc=featured_projects
3. Greenspan, A. L. 2004. *Employer's guide to workplace privacy.* New York: Aspen Publishers.

Chapter 10

Keys, Locks, and Safes

Paul R. Baker and Daniel J. Benny

Keys, Locks, and Safes

According to Northern Virginia Lock and Security Systems, a lock is defined as "a device that incorporates a bolt, cam, shackle or switch to secure an object—such as a door, drawer or machine—to a closed, opened, locked, off or on position, and that provides a restricted means of releasing the object from that position" (personal communication with Eric Smith, owner of NoVA Lock and Security, June 2, 2011).

The use of locks is one of the oldest forms of security and is still utilized in current and new construction building today. There are two general categories of locks: those that operate on mechanical concepts and those that use electricity to operate mechanical components of the locking system. In addition to preventing access based on security concerns, locks can also prevent access to areas for safety-related issues. This might include securing hazardous materials storage areas, electrical rooms, and locking out equipment on/off switches.

Key locks are one of the basic safeguards in protecting buildings, personnel, and property and are generally used to secure doors and windows. According to UL 437, door locks and locking cylinders must resist attack through the following testing procedures: the picking test, impression test (a lock is surreptitiously opened by making an impression of the key with a key blank of some malleable material, such as wax or plastic, which is inserted into the keyway and then filed to fit the lock), forcing test, and salt spray corrosion test for products intended for outdoor use. The door locks and locking cylinders are required by UL standards to resist picking and impression for 10 minutes.

Locks

Mechanical lock. A mechanical lock utilizes physical moving parts and barriers to prevent the opening of the latch and includes the following:
- The latch or bolt that holds the door or window to the frame
- The strike into which the latch is inserted
- The barrier—a tumbler array that must be passed by use of a key to operate the latch
- The key that is used to pass through the tumbler array and operate the latch or bolt

Rim lock. A rim lock, shown in Figure 10.1, is a lock or latch typically mounted on the surface of a door. It is typically associated with a deadbolt type of lock.

Mortise lock. A mortise lock (Figure 10.2) is a lock or latch that is recessed into the edge of a door, rather than being mounted to its surface. This configuration has a handle and locking device all in one package.

Locking cylinder. The pin tumbler cylinder is a locking cylinder that is composed of circular pin tumblers that fit into matching circular holes on two internal parts of the lock (Figure 10.3). The pin tumbler functions on the principle that the pin tumblers need to be placed into a

Figure 10.1 A rim lock is a lock or latch typically mounted on the surface of a door.

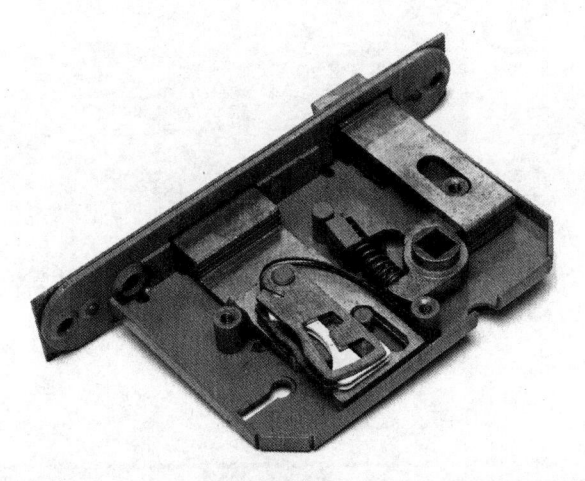

Figure 10.2 A mortise lock is a lock or latch that is recessed into the edge of a door, rather than being mounted to its surface.

position that is entirely contained with the plug. Each pin is of a different height, thus accounting for the varying ridge sizes of the key. When the pins are properly aligned, the plug can be turned to unlock the bolt.

Cipher lock. A cipher lock, shown in Figure 10.4, is controlled by a mechanical key pad, typically five digits that, when you push in the right combination, the lock will release and allow entry. The drawback is that someone looking over your shoulder can see the combination. However, an electric version of the cipher lock is in production in which a display screen will automatically move the numbers around, so

Figure 10.3 A pin tumbler cylinder is a locking cylinder that is composed of circular pin tumblers that fit into matching circular holes on two internal parts of the lock. (Courtesy of Bosch.)

Figure 10.4 A cipher lock is controlled by a mechanical key pad, typically five digits that, when you push in the right combination, the lock will release and allow entry.

that if someone is trying to watch your movement on the screen, he or she will not be able to identify the number you are indicating unless he or she is standing right behind you.

Remember that locking devices are only as good as the wall or door that they are mounted in. If the frame of the door or the door itself can be easily destroyed, then the lock cannot be effective. A lock will eventually be defeated and its primary purpose is to delay the attacker.

High-Tech Keys

Not all lock and key systems are your standard metal composite. There have been developments in key technology that offer convenient, reliable access control.

Intellikey (manufactured by Intelligent Access Systems, Atlanta, Georgia) is a key with a built-in microprocessor; it is unique to the individual key holder and identifies the key holder specifically. The lock, which also contains a minicomputer and the key exchange data, allows the lock to make valid access decisions based on the parameters established for the key holder. For example, the key will know if the employee is allowed access into the facility after normal business hours; if this is not the case, the key will not work. Also, it will keep track of whose key is being used to access specific locked doors, and when the attempts are taking place. When an employee resigns from the organization, the relevant key is disabled.

Instakey (manufactured by Shield Security Systems, Englewood, Colorado) provides a quick way to disable a key by permitting one turn of the master key to change a lock. This method of changing a lock can save both time and money in the event a master key is lost. According to the manufacturer, a 50-story bank building can be rekeyed in 6 hours by two security guards. The Instakey system can go through 10 to 15 changes before having to be repinned.

Key Control

Key control or, more accurately, the lack of key control is one of the biggest risks that businesses or property owners face. Strong locks and stronger key control are the two essentials in a high-security locking system.

In most cases master and submaster keys are required for most building systems so that janitorial and other maintenance personnel may have access. Thus, the control of all keys becomes a critical element of the key lock system. All keys need to be tightly controlled from the day of purchase by designated personnel responsible for the lock system.

Without a key control system, you cannot be sure who has keys or how many keys to your property someone has. Not having a patent-controlled key system leads to unauthorized key duplication, which leads to unauthorized access to your property or to employee theft.

Developing a Master Locking System

When establishing a master locking system, it must be designed to meet the security needs of the organization. Without planning, the locking system will usually degrade to a system that is only providing privacy but not effective security. The goal is to make the locking system effective and user friendly so that the functions of the organization can continue unimpeded.

The following design criteria need to be considered in the development of a master locking system:

Number of locks. This includes the total number of locks that will be installed in the facility on exterior and interior doors.
Categories of the locking system. The categories of a locking system would include exterior gates on the perimeter of the property, exterior doors entering the building on the property, interior doors, high-security

areas, combination locks for security containers and safes, and desk, computer, and file cabinet locks.

Security objectives. When securing a facility utilizing locks, it is important to determine the security objectives of the areas being secured. Perimeter doors and doors leading into high-risk or high-value areas such as the president's office, security office, cash office, or computer rooms would be considered a high-security area. Secondary areas would include locations in the facility where it is important to restrict access, but where the threat or risk is not as high.

Workforce size and turnover. It is vital to preserve the integrity of the locking system. The more individuals participating in the system, the better chance there is of lost keys and compromised locking devices.

Most key control systems utilize patented keys and/or cylinders. These lock cylinders employ very precise locking systems that can only be operated by the unique keys to that system. Because the cylinders and/or the keys are patented, the duplication of keys can only be done by factory-authorized professional locksmiths.

The key blanks and lock cylinders are made available only to those same factory-authorized professional locksmiths. Procedures may be in place to allow you to contract with another security professional, should the need arise.

All high-security key control systems require specific permission to have keys originated or duplicated. These procedures assure the property owner or manager that he or she will always know who has keys and how many. If an employee leaves and you get the keys back, you can be reasonably assured that no copies of your keys are floating around.

Most systems have cylinders that will retrofit existing hardware, keeping the cost of acquisition lower. Some systems employ different levels of security within the system—still giving patented control, but not requiring ultrahigh security where it is not needed. These measures are again aimed at cost control.

Most systems can be master keyed; some will coordinate with existing master key systems. There are systems available that allow interchangeable core cylinders that will retrofit existing interchangeable core systems.

Locks, keys, doors, and frame construction are interconnected and all must be equally effective. If any one link is weak, the system will break down.

The following is from the *Medeco Guide for Developing and Managing Key Control:*[1]

1. This facility shall use a key control system and adopt administrative policies that facilitate the enforcement of Key Management Procedures.
2. The following represent the basic and most critical elements of key control and shall be included, as a minimum, in the key control specification:
    2.1 Facility shall appoint a Key Control Authority or Key Control Manager to implement, execute, and enforce key control policies and procedures.
    2.2 A policy and method for the issuing and collecting of all keys shall be implemented.
    2.3 Keys and key blanks shall be stored in a locked cabinet or container, in a secured area.
    2.4 A key control management program shall be utilized. A dedicated computer software application is preferred.
    2.5 All keys shall remain the property of the issuing facility.
    2.6 A key should be issued only to individuals who have a legitimate and official requirement for the key.
        2.6.1 A requirement for access alone, when access can be accomplished by other means (such as unlocked doors, request for entry, intercoms, timers, etc.), shall not convey automatic entitlement to a key.
    2.7 All keys shall be returned and accounted for.
    2.8 Employees must ensure that keys are safeguarded and properly used.

Safes

Within the inner sanctum of the security envelope, secured rooms are a necessity. To further the security of information, there is a need for vaults, safes, and containers.

A safe (Figure 10.5) is defined as a fireproof and burglarproof iron or steel chest used for the storage of currency, negotiable securities, and similar valuables. The categories for safes depend on the amount of security needed. Underwriters Laboratories lists several classifications of safe; the following is one such classification:

Figure 10.5 A safe is a fireproof and burglarproof iron or steel chest used for the storage of currency, negotiable securities, and similar valuables.

- **Tool-resistant safe class TL-15.** This type of combination lock safe is designed to meet the following requirements: It must be resistant to entry (by opening the door or making a 6-inch hand hole through the door) for a net working time of 15 minutes using any combination of the following tools: mechanical or portable electric hand drills not exceeding 1/2-inch size, grinding points, carbide drills (excluding the magnetic drill press and other pressure-applying mechanisms, abrasive wheels, and rotating saws), and common hand tools such as chisels, drifts, wrenches, screwdrivers, pliers, and hammers and sledges not to exceed the 8-pound size, pry bars and ripping tools not to exceed five feet in length, and picking tools that are not specially designed for use against a special make of safe. A TL-15 safe must:
    - Weigh at least 750 pounds or be equipped with anchors and instructions for anchoring in larger safes, in concrete blocks, or to the floor of the bank premises.

- Have metal in the body that is solid cast or fabricated open-hearth steel at least 1 inch thick with a tensile strength of 50,000 pounds per square inch (psi) and that is fastened to the floor in a manner equal to a continuous 1/4-inch penetration weld of open-hearth steel having an ultimate tensile strength of 50,000 psi.
- Have the hole to permit insertion of electrical conductors for alarm devices not exceed a 1/4-inch diameter and be provided in the top, side, bottom, or back of the safe body, but must not permit a direct view of the door or locking mechanism.
- Be equipped with a combination lock meeting UL Standard No. 768 requirements for Group 2, 1, or 1R locks.
- Be equipped with a relocking device that will effectively lock the door if the combination lock is punched.

The UL classifications mean that a tool-resistant safe class TL-30 will take 30 minutes, using tools, to break into the safe. A TRTL-30 safe means it will take 30 minutes for a combination of tools and torches to break into the safe. The categories go up to a safe that can resist tools, torches, and explosives.

Vaults

A vault is defined as a room or compartment designed for the storage and safekeeping of valuables and has a size and shape that permit entrance and movement within by one or more persons. Vaults generally are constructed to withstand the best efforts of man and nature to penetrate them.

The UL has developed standards for vault doors and vault modular panels for use in the construction of vault floors, walls, and ceilings. The standards are intended to establish the burglary-resistant rating of vault doors and modular vault panels according to the length of time they withstand attack by common mechanical tools, electric tools, cutting torches, or any combination thereof. The ratings, based on the net working time to effect entry, are as follows:

- Class M—¼ hour
- Class 1—½ hour
- Class 2—1 hour
- Class 3—2 hours

Containers

A container is a reinforced filing cabinet that can be used to store proprietary and sensitive information. The standards for containers are from the government, which lists a class 6 container as GSA approved for the storage of secret, top secret, and confidential information. The container must meet the protection requirements for 30 man-minutes against covert entry and 20 man-hours against surreptitious entry with no forced entry.

Reference

1. http://www.medeco.com/techsvc/keyControlGuide.html

Chapter 11

Biometrics

Roque Solis and Chuck Wilson

Introduction

The underlying purpose of identity management in physical security is to verify the identity of individuals in order to provide access to those who are authorized and to deny access to those who are not authorized. Nearly all identification methods in use today are based on one or more tokens or identifiers:

- There are knowledge tokens such as passwords or PINs or relatively obscure personal information such as one's mother's maiden name.
- There are physical tokens such as badges, cards, passports, driver's licenses, and keys.
- There are behavioral or physiological biometric identifiers.

Each of these methods has its advantages and limitations, and each may be the appropriate solution in a given set of circumstances. But each method by itself offers only a single security factor. Strong authentication[1] demands a combination of authentication factors such as knowledge, tokens, or biometric authentication used to verify identity under security constraints.

Although tokens can provide a proxy for an identity claim, they can be forgotten, lost, stolen, forged, or compromised, and they do not provide nonrepudiation.[2] Of course, some card tokens can securely store passwords or PINs, combining two separate security factors. Their joint use significantly improves the likelihood that they are legitimate; yet even together they

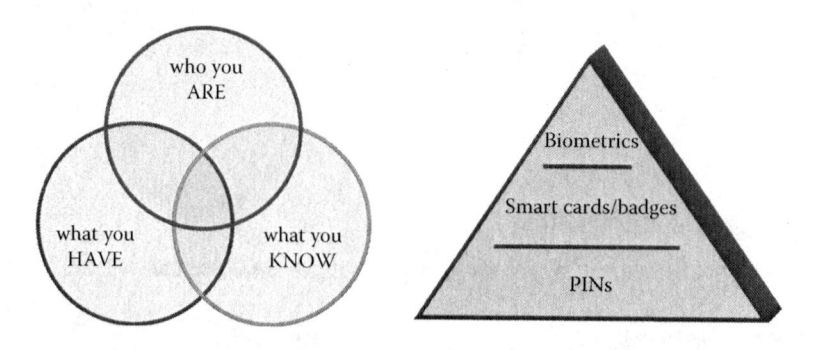

Figure 11.1 Relationship of three-factor security.

remain surrogate representations to authenticate one's identity. Token-based systems simply do not completely verify the identity of the user.

With regard to physical security, in order to enable entry by authorized individuals and to thwart the unauthorized ones, one's identity must be verified within an acceptable probability. Biometrics is arguably the only technology that can bind a person to a verification event,[3] such as requesting entry into a secure facility. A biometric template could also be stored on a PIN-activated smart card (which might also serve as a badge) and, together, they would provide three-factor security. When strong three-factor security is used to secure physical access, the risk of fraud significantly declines, and assurance of legitimacy substantially increases. Figure 11.1 illustrates the relative power of triple factor security.

What Is Biometrics?

Biometrics means "life measurement." Biometrics refers to automated methods of recognizing an individual based on a measurement of his or her physical or behavioral characteristics. A biometric system is basically an automated pattern recognition system that either makes an identification from a database of many (this is referred to as 1:N, and it answers the question, "Who are you?") or verifies a professed identity (this is referred to as 1:1, and it answers the question, "Are you who you claim to be?"). Most nonsurveillance, physical security applications employ identity verification methods.

Based on the use of unique and measurable characteristics, biometrics validates a professed identity, within a given probability. Physiological characteristics include, but are not limited to, a person's vein patterns, facial structure, ocular characteristics, hand geometry, or fingerprints. Behavioral characteristics are based on an individual's unique actions captured over a period of time, including such traits as signature dynamics, voice, gait, and keystroke dynamics. Biometrics is considered a natural and reliable solution for identity verification situations. Therefore, the inclusion of a biometric component for identity verification is a critical enhancement for many physical security systems.

There are a variety of biometric systems in use today. However, no single biometric type, referred to as the biometric modality, is best for all situations. The right biometric modality for a given application depends on many factors, including the task at hand, security risks, population of users, and user circumstances. Moreover, biometric modalities are in varying stages of maturity in terms of implementation experiences. So a key question might be, "What is the right biometric identifier for a given application?"

Obviously, that would depend on the evaluation criteria one uses. If one were addressing a telephonic application such as a Help Desk situation, then speaker recognition might be used. For a financial application that is legally binding, dynamic signature capture would seem appropriate. For physical security, a variety of biometric modalities come to mind, but given that we have limited pages to address biometric solutions, we have chosen the five most popular physical access modalities in use today: (1) facial recognition, (2) fingerprint/palm print, (3) hand geometry, (4) iris recognition, and (5) vein pattern recognition.

Facial Recognition

Facial recognition is the most natural means of identity recognition. Facial recognition systems attempt to find recognizable facial characteristics, reduce the key features to digital representations, and match them against known facial templates. Although various techniques can be used, facial recognition systems tend to emphasize facial regions that are less susceptible to alteration, such as the areas surrounding the cheekbones, upper outlines of eye sockets, the distance between the eyes, and the sides of the mouth.

Facial recognition systems can process a two-dimensional or three-dimensional camera image, depending on the system (Figure 11.2). Face scanning can be accomplished at a distance for clandestine surveillance or in close proximity to the subject. Two-dimensional facial recognition systems have not achieved high accuracy relative to some other biometric techniques, but they are constantly improving. Three-dimensional facial

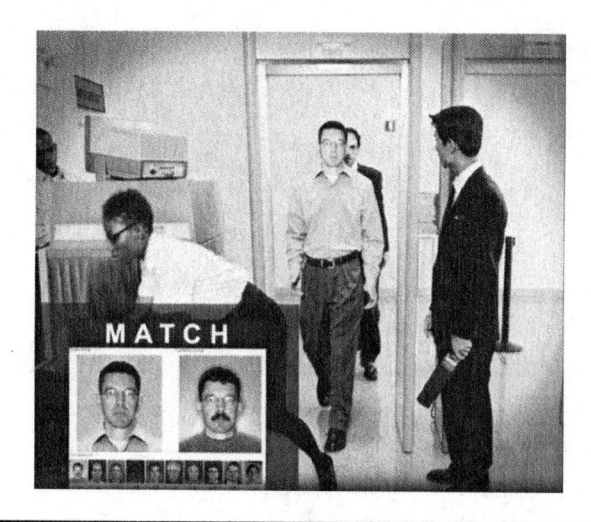

Figure 11.2 Physical access application for facial recognition. (Courtesy of L-1 Identity Solutions.)

recognition is not as dependent on lighting conditions or varied poses, including nonfrontal angles. Three-dimensional systems are still developing, but they have the potential for high accuracy.

The advantages of facial recognition include high public acceptance of the modality, commonly available sensors (e.g., cameras), not physically intrusive, and the ease with which humans can verify the results. A key advantage to facial recognition is that this biometric identifier can be used at any venue in which one can mount a camera. An interesting by-product of this biometric is that the system administrator can retrieve a visual record of unauthorized attempts at entry. The disadvantages include its susceptibility to "spoofing"[4] in large databases, as the face can be obstructed; its sensitivity to lighting, facial expressions, and pose; its large templates (approximately 3,000 bytes or more); and the reality that faces do change with age.

Facial recognition has been used successfully throughout gaming casinos since the late 1990s as a mechanism to identify banned gamblers. Facial recognition has been a well-used biometric identifier for physical access applications.

Fingerprint

Fingerprinting is the oldest biometric method in use today; due to its significant head start, it is the most widely used biometric modality. Fingerprints are based on physical dermal structures defined before birth. The pattern of furrows and ridges on the surface of each finger (e.g., local ridge characteristics that occur at a ridge bifurcation or a ridge ending) and image density

Figure 11.3   Fingerprint.

are unique to each individual. The print pattern of a finger's top joint (the distal interphalangeal joint) is the primary focus, as shown in Figure 11.3.

Fingerprint-based systems work by scanning the tips of one or more fingers and comparing the scans of the finger against known images—the pre-established fingerprint template created during program enrollment. Traditional fingerprint sensing systems use a variety of techniques to capture a fingerprint image including capacitive,[5] ultrasound,[6] and optical[7] methods.

Fingerprint (FP) scanning has several advantages. Given its history and common use, most people understand the concept, and training is minimized. Figure 11.4 depicts a fingerprint system mounted on a wall adjacent to a secured entry point. The sensors used in fingerprint applications tend to be quite small, allowing some very useful form factors for FP scanners. The FP templates are also small, enabling a greater volume of template storage in limited memory situations, as well as enabling a very fast retrieval process. Most types of FP scanning equipment are relatively inexpensive.

Fingerprint scanning does have an extensive set of limitations that should be considered. Biometric industry authorities indicate that between 2% and

Figure 11.4   Worker gaining physical access. (Courtesy of L-1 Identity Solutions.)

5% of the general population has some physical limitation in regard to using fingerprint imaging technology. Fingerprint ridges deteriorate with age and wear. Some occupational activities, such as aggressively washing one's hands multiple times a day (e.g., surgeons) or constantly placing one's fingers in strong chemicals (e.g., hair beauticians), can erode one's fingerprints over time.

Depending on an individual's environment, his or her fingertips may rather easily become dirty, oily, or cut. Depending on the type of sensor being used, dirt and oil could obscure the finger image, as could cuts, calluses, or scars on the fingertip. Fingerprint templates might be significantly influenced by dryness of the fingers or by the pressure or positioning of the fingers for a scan.

It is well known that fingerprints have a latency property. That is, we leave our fingerprints behind on just about every surface we touch: from the drinking glass we hold at restaurants to the doors we open. This can impact privacy and can encourage imposters to try to spoof the system. Moreover, in some countries there is a cultural association of fingerprinting with forensics and the criminal element of society. However, several emerging techniques show great promise in alleviating some of the traditional shortcomings associated with fingerprints; these techniques include multispectral imaging (MSI),[8] 3D contactless finger imaging,[9] and 3D ultrasonic imaging.[10]

Very similar to fingerprint recognition, palm print recognition measures the prints in one's palm to verify identity. Palm print recognition also has a similar set of advantages and limitations. Since the palm is larger, more distinctive features can be acquired in comparison with fingerprints. On the other hand, palm print scanners tend to be bulkier than fingerprint scanners and are usually more expensive. Many commercial fingerprint system providers have added palm print scanning to their biometric lineup of products.

Hand Geometry

Biometric systems based on hand geometry focus on the physical structure (lengths, widths, thicknesses, and angles of the fingers and palm) of an outstretched hand. The human hand has 27 bones, and it provides adequate anatomical features for measurement to enable verification. Typically, a user places his or her hand on a reader that is studded with guide pegs. The advantages of hand geometry are that it is intuitively easy to use, it typically uses 9–20 bytes of data for its templates (the smallest template of any biometric modality), its error rates are very low for false reject and failure to enroll, and it has fewer privacy concerns than fingerprint or face recognition.

Hand geometry devices can function in extreme conditions, withstand wide temperature variances, and are not impacted by dirty hands or dusty environments. The modality is well suited for industrial locations such as warehousing or some manufacturing facilities (for physical access control and time and attendance applications), where its simplicity is a positive. Moreover, hand geometry can be readily combined with other biometrics, such as fingerprints and vein pattern recognition of the hand or fingers.

For better or worse, the human hand is not truly unique, causing hand geometry scanners to operate with a relatively high false-accept rate (on the order of 0.1%). Additionally, the features of one's hand can vary over time (more than most biometric modalities). The primary disadvantages of hand geometry include the relatively large size of the readers, its moderate accuracy rates, and the sanitation concerns of placing one's hands in the same place as dozens or even hundreds of other people that preceded the user. Current research work involves identifying new features that would result in better discrimination among different hands.

Modern hand geometry devices have been successfully manufactured since the early 1980s, placing hand geometry among the first biometric modalities to find widespread use. Over 300,000 hand geometry readers have been sold worldwide, including to the San Francisco International Airport, US nuclear power plants, military bases, universities, day care centers, welfare agencies, and hospitals. They are also used to streamline security and immigration procedures for over 90,000 users[11] in the INSPASS[12] frequent international traveler system.

Iris Recognition

Iris recognition systems use small, high-quality, near-infrared cameras to capture a black and white, high-resolution, high-contrast photograph of the iris. The near infrared illumination reveals patterns even for dark irises with no discernible patterns during visible light. Once the image is captured, the iris's elastic connective tissue is analyzed, and the distinctive features are extracted and translated into a digital form. Iris-based systems are relatively nonintrusive and are hygienic.

The iris is the annular region of the eye bounded by pupil at its center and by the sclera (white of the eye) on its outer side (Figure 11.5). Every iris is distinct! An iris is highly stable over time, and it is less subject to wear or injury than most other measurable body parts. The iris offers a data-rich structure composed of fibrous and vascular tissue including freckles,

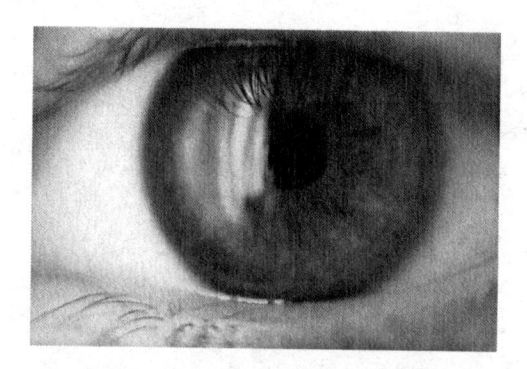

Figure 11.5 Photo of an iris.

furrows, pits, rings, corona, and striations—all of which can be measured and the intraspatial relationships gauged. The detail in the texture of the iris is established before birth, and it appears to be caused by random processes, referred to as chaotic morphogenesis.[13]

Iris scanners can measure the characteristics in the iris at a distance of a meter or more. The muscles in the iris control the amount of light entering the pupil and thus regulate the size of the pupil. The iris reacts to the stimulus of light with involuntary reflexes throughout the iris muscles. Some iris systems use this in a challenge/response effort to guarantee a living eye and not an artificial image. Iris scanning was specifically designed for physical access control applications.

The collection of an iris image does require more training and attentiveness than most biometric modalities. Poor-quality images can result in higher than normal failure to enroll rates in some systems. Most iris recognition systems are somewhat confining based on the required read distance,[14] the need for special lighting, and the desired camera angle (position of the eye in relation to the camera). Also, iris scanning systems tend to be among the most expensive biometrics.

Vein Pattern Recognition

Based on the unique patterns of veins in one's finger or hand, vein pattern recognition (VPR),[15] also referred to as vascular pattern recognition,[16] provides ease of use, a high level of accuracy, small readers, and a secure and hygienic approach (Figure 11.6). It uses near-infrared light[17] generated from a bank of LEDs projected through an individual's skin to enable a high-contrast matching of vein patterns (e.g., vein branching points, branching angles, etc.) in one's hand or finger.

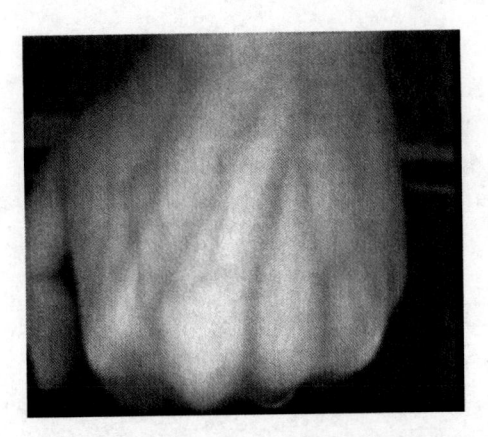

Figure 11.6 Vein pattern recognition. (Courtesy of Identica Corp.)

VPR systems scan the deoxygenated veins in one's palm, back of the hand, or fingers; extract key pattern features via contactless near-infrared optical sensor systems; digitize the extracted pattern recognition; and then match the transaction templates to the respective pre-established enrollment template. By measuring the veins under the skin with near-infrared spectroscopic imaging, it is very difficult for unauthorized persons to observe or capture this pattern, rendering VPR a highly secure method of identity verification.

VPR systems have some very powerful advantages. Vein pattern sensors look below the skin and they generally do not have issues with minor cuts and moisture. VPR systems reveal no significant performance degradation when measuring sweating or mildly dirty hands. They demonstrate very high accuracy rates and they are very difficult to spoof since blood needs to flow to register an image. Only a slight portion of the hand makes contact with the scanning device or its guideposts in order to align the finger or hand for consistent imaging. Since users do not touch the sensor surface and some vein pattern scanners are totally contactless, they are considered hygienic. VPR systems are easy to use and require only a moderate level of training on the part of the user. VPR systems have demonstrated exceptionally high usability rates exceeding 99.9% of most populations.[18]

As with all biometrics, VPR systems have their limitations. They can be impacted by direct sunlight or other ambient bright light; that is why some VPR systems have been positioned primarily for indoor use. However, many VPR system providers have designed coverings for outside access systems; some even have semitransparent coverings to limit ambient light indoors.

Multibiometric Systems

Security systems are not perfect; neither are biometric solutions. That is why the concept of multiple layering continues to gain traction. A multisystem security strategy is basically the use of two or more levels or types of security techniques. The concept is to use multiple techniques or technologies in a layered approach. Multiple security systems used together are generally more resistant to fraud since they employ different techniques or technologies and process information through different algorithms. Similarly, using a multibiometric system can increase security (e.g., by improving accuracy) or broaden support for and acceptance by the user population by offering alternatives. Multiple biometric techniques can be integrated into one security system to achieve better the stringent performance requirements imposed by some high-security applications, to identify imposters more readily, and to decrease the opportunities for imposters to spoof the system.

The concept of multibiometric systems has been used commercially since the late 1990s. A multibiometric system is one that uses multiple biometric sources to improve its overall performance. As illustrated in Figure 11.7, a

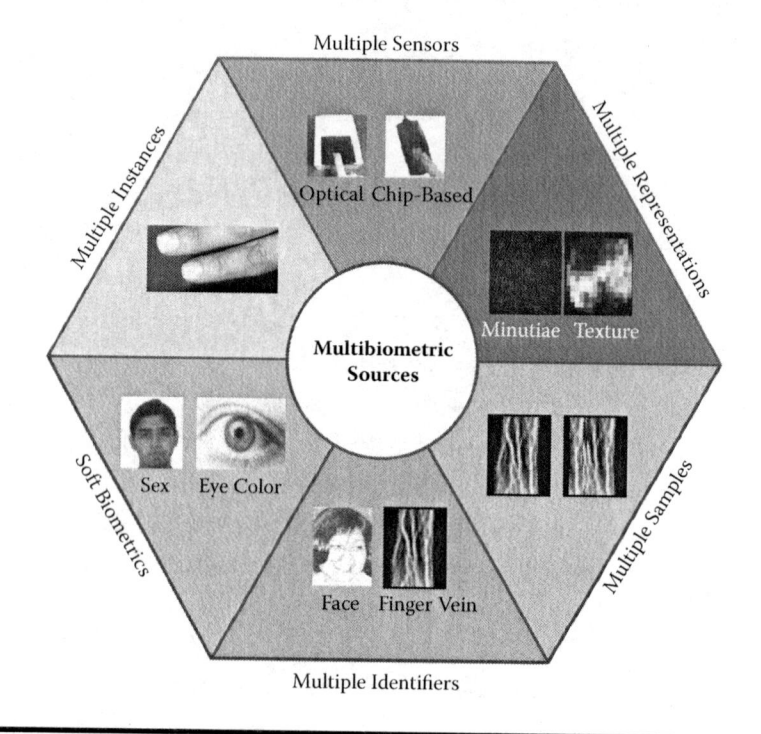

Figure 11.7 Multibiometric sources.

multibiometric system might use any of the following six scenarios (starting at the top and continuing clockwise):

1. Multiple sensors producing multiple samples of a single biometric trait to assuage noisy sensor data so that different sensors (e.g., optical and chip-based sensors for fingerprinting) for the same biometric identifier might be used to improve performance
2. Multiple representations from the same biometric trait (e.g., minutiae-based vs. filter-based fingerprint systems) involving different approaches to feature extraction and matching
3. Multiple samples of the same biometric trait using the same instance (e.g., same finger or same hand)
4. Multiple identifiers, usually referred to as multimodal biometrics (e.g., using fingerprint and finger vein pattern recognition). Multimodal systems are arguably the most powerful type of multibiometric system. Indeed, multimodal systems analyze the evidence from multiple sources for verifying an individual's identity or for identifying an individual from a database, and generally multimodal systems provide superior recognition performance over unimodal biometric systems.

    Admittedly, multimodal systems usually require more time and effort for users to enroll and to verify themselves. They also introduce a new level of complexity. Due to considerations of information integration, total cost, and end-user training, past successes with multimodal systems have been limited primarily to high-security facilities where cost issues are not a prime consideration and where users are motivated to master these systems.
5. Using "soft biometrics" such as gender, height, or eye color as a way to strengthen performance and accelerate processing time
6. Sampling multiple instances within the same modality (also called intramodal) such as different fingers or both palms for vein pattern recognition, which can help avoid spoofing in a challenge/response authentication

The choice of multiple integration strategies depends primarily on an enterprise's requirements as well as the types of applications supported, the correlations among the biometric identifiers, and, of course, the costs incurred. These techniques provide multiple corridors of security checks that can be performed simultaneously or sequentially. Multiple biometric techniques combine multiple factors of evidence to enable better decisions.

Deploying Biometrics

Prior to deploying a biometric system, one must plan how the system will be used and must render some decisions on the best way to effect template storage, proper enrollment techniques, and daily biometric transaction use. And one should always build in privacy-enhancing features from the beginning.

Enrollment registers an individual to a biometric database for the first time. This is the process in which a user presents a biometric sample for the system to convert to a reference template and to store it in the biometric system database along with a proxy for his or her identity. The user's initial biometric sample is collected, assessed, processed, and stored. To a large degree, enrollment helps determine the ultimate accuracy of future matches.

Generally, no two reads from any biometric reader are exactly the same, and thus no two templates from the same person are *exactly* the same. Small variations in one's positioning, the distance from the sensor to an individual's identifying feature, the subject's interface with the sensor, and environmental conditions (humidity, temperature, sunlight, etc.) contribute to rendering a template unique. Each time biometric data are extracted, they are used to create a unique sample template for comparison to the reference template. The quality of each template is critical to the success of the overall system. The better the quality of the templates is, the more accurate will be the overall system and the lower the error rate.

Enrollment is the most crucial stage of the biometric process. Nothing can influence the successful use of the biometric system more than a proper enrollment. If one fails to capture the biometric trait properly and create an accurate reference template, then subsequent attempts to match templates will fail, and that will continue through the biometric life cycle for that individual. Thus, a weak enrollment process may lead to system inaccuracies and inconsistencies, as well as inadvertently creating an unreliable authentication infrastructure.

During the enrollment process a user record is created. That user record consists of an identifier and the biometric template, both of which should be encrypted (Figure 11.8). The identifier links the template and the user. A single user can be associated with various identifiers, which may lead to multiple roles in the biometric system for the attribution of rights and privileges. This also prevents an attacker from taking someone else's user credential by attempting to link his identifier to it.

During enrollment, most biometric systems collect ancillary information about the enrollees. This ancillary information may include demographic

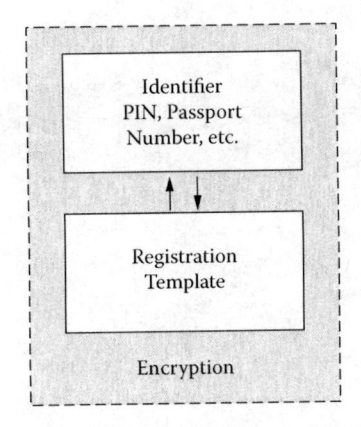

```
┌─────────────────────────────┐
│   ┌─────────────────────┐   │
│   │     Identifier      │   │
│   │   PIN, Passport     │   │
│   │    Number, etc.     │   │
│   └─────────────────────┘   │
│            ↑ ↓              │
│   ┌─────────────────────┐   │
│   │    Registration     │   │
│   │      Template       │   │
│   └─────────────────────┘   │
│         Encryption          │
└─────────────────────────────┘
```

Figure 11.8 User record.

information (e.g., organization, division, address) or soft biometric information (e.g., gender, age, eye color) to help in verifying identity. It is very important for the organization collecting this information to store and safeguard it securely, partition it for specific uses, and delete it when the individual is no longer participating in that particular endeavor.

Each enrollee usually presents some background information that vouches for his or her identity, but fraudulent information such as a fake or stolen birth certificate or driver's license might be presented. For this reason supervisors of the enrollment process must insist on multiple valid methods of identity proof. To help ensure the integrity of the enrollment process, one might perform a matching process of each new enrollment against the existing database to check for duplicate entries.

A **verification transaction** is any application that requires individuals to authenticate a claimed identity. Moreover, verification activities tend to be self-service based, transaction oriented, and unsupervised. The biometric system captures an individual's biometric image and then extracts the unique characteristics from the individual's image to create the user's sample template. The biometric system then compares the sample template to the template stored at enrollment (e.g., the reference template), and in most systems, a numeric matching score is generated based on the percentage of similarity between the sample and reference templates.

Depending on the predetermined threshold value, the identity verification score may or may not meet the probability threshold for a match. Comparison of sample data and reference templates results in a biometric system match decision—that is, a yes or no match. Figure 11.9 illustrates a standard biometric verification process.

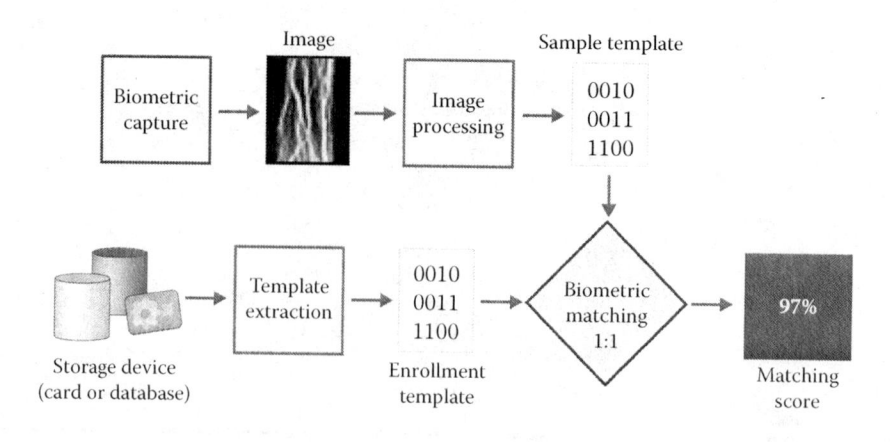

Figure 11.9 Verification process.

Verification tasks are considered to be relatively simple and straightforward. The entire process for a verification application generally takes only a few seconds, depending on the biometric modality employed and the specific implementation. Some systems allow multiple sample retakes to produce a match. If an individual is denied, human interaction may be needed, and the denied person may need to follow a manual process for authentication.

Template storage houses the enrollment templates to which the new biometric templates will be compared. Templates are stored within an enrollment database held in the data storage component. The templates can be stored in an altered format, compressed, and encrypted.

Since templates require only modest memory, ranging from less than 15 bytes (e.g., hand geometry) to around 3,000 bytes (e.g., facial or speaker recognition) in size, storage space is not typically a major issue, except in very large implementations. There are typically three options for reference template storage for physical security systems:

- Locally store the template within the biometric reader itself or in another localized database. This enables a fast response during future verification transactions. However, whenever individuals need to access multiple locations across a given geography, the database must be replicated at each reader. This may require extensive memory within the reader, and it imposes the requirement to update each reader frequently.
- Remotely store the template in a central data repository. This generally works well for identification applications where sample templates

must be compared to the entire template repository. However, for distributed verification applications, access to a central template repository would heavily depend on the data network that connects it to the reader device and it would introduce multiple points of failure. It would be vitally important that all transmitted data be encrypted to counter fraudsters who might otherwise "sniff" the biometric data off the network and replay the authentication session in an attack. In addition, some users are very privacy conscious and do not like the idea of their enrollment templates being stored centrally.

- Securely store the template on a portable token such as a smart card, such as the one shown in Figure 11.10. With this method, individuals carry their biometric templates with them on their smart cards. Matching algorithms can be implemented directly via on-card matching. In this way, the enrolled biometric template never leaves the card, and the card remains with the individual. The one drawback is that the cost of the biometric implementation is marginally higher, since smart cards and smart card readers would also be required. However, both cards and card readers are relatively inexpensive. Indeed, new smart card form factors have been developed such as USB-based smart tokens for which no external reader is required. In most applications smart cards are the container of choice for biometric templates because they offer greater mobility, flexibility, and security; have fewer points of failure; and allow for better support of privacy goals.

Figure 11.10 Smart card readers can be used for both physical as well as computer systems access.

Privacy

Although biometric technology can enhance and support most commonly held notions of privacy, it cannot guarantee it. Clearly, the topic of biometrics often raises the public's angst regarding privacy. There has long existed within some groups an inherent discomfort with biometrics; they have a perception that biometrics represents privacy intrusiveness. As we have mentioned previously, we leave our fingerprints on everything we touch, and there are limited prohibitions against someone taking our photographs or recording our voices, each of which can be accomplished legally without our awareness or consent, depending on the circumstances.

At the vortex of most privacy issues are concerns about the potential misuse of biometrics, such as a biometric system pointing to other information pertaining to an individual. Thus, an intrusion on one's privacy tends to originate from the inherent misuse of the technology, not from the biometric application itself. Identity verification applications may be less threatening to privacy (than, say, surveillance applications) simply because the individual knowingly enrolls in a given system and thereby gives permission to use his or her biometric characteristics for a specific purpose. However, that information can still be misused.

It is generally the case that where there is biometric information, there is probably some personal information. An enterprise planning to use biometrics should conduct a privacy assessment to analyze what personal information is being collected and the necessity for doing so, as well as how that personal information will be used.

Biometrics can be privacy threatening if safeguards are not designed into biometric systems from the beginning. It really depends on how the biometric system is deployed, how the biometric templates are stored, and the strength and vigilance of the safeguards that are put in place to guard against third-party access or linking inappropriately to other systems.

Biometric Metrics

The most commonly used metrics for biometric evaluation are failure to enroll rate/failure to acquire rate (FTER/FTAR), false acceptance rate (FAR), and false reject rate (FRR).

Failure to Enroll

Given a large population of users, regardless of the technology, some small percentage will be unable to enroll since no biometric modality can claim to work for 100% of a sizeable population.

A biometric system's FTE is the rate at which potential users are rejected from enrollment in a biometric system due to insufficiently distinctive biometric samples or their inability to interact with the system. In any population of users, there will be individuals whose physical attributes militate against successful enrollments. For example, an individual with a musculoskeletal disorder might be unable to provide his fingerprint image because he cannot place his finger on the appropriate sensor. Other reasons for failure to enroll include inadequate user training, poor supervisor training, device time-outs, or the inability of a system to sense or locate a presented sample. The failure to enroll rate (FTER) is the proportion of users for which a biometric system cannot generate acceptable reference templates.

What happens if a person cannot be enrolled? It is important for the enterprise to plan for and implement alternatives for these individuals. This is referred to as exception handling, and all biometric systems will have an occasional need for it. From a systems integrity and fairness standpoint, the exception handling should always be consistent, thorough, and each situation well documented.

Failure to Acquire

FTA is similar to FTE except that it occurs during verification transactions as opposed to enrollment transactions. FTAR is the rate at which a biometric system fails to capture or extract *usable* information from a biometric sample. As used here, *usable* refers to an image or signal of sufficient quality to enable a comparison. This could occur for a whole range of reasons including equipment malfunction, interference with the image sensor (e.g., dirt particles on the camera), environmental issues, or human anomalies such as improper alignment of the user's finger or hand, or even adjustable thresholds for image or signal quality. Generally, when a biometric system allows multiple attempts, FTA measures the biometric system's failure to capture throughout these multiple attempts.

FRR/FAR

One of the metrics used for verification systems is a false rejection rate (FRR), which is an unintended reject, and thus a FRR measures the portion of genuine users that are rejected. This occurs when an authorized individual is not correctly matched to his or her own existing reference template. For a verification system, it can be estimated as

(the number of false rejects)/(the number of verification attempts).

In Figure 11.11, a user approaches a biometric system and claims to be Jose Gonzalez. He submits his hand to the biometric reader, which creates a template to match against the stored template from enrollment. The verification threshold is set at 95, and Mr. Gonzalez's biometric score is 97.5 and deemed a match. We can safely say that Mr. Gonzalez is who he says he is. However, on one particular day the FV system calculates a similarity score[19] for Mr. Gonzalez at 93.7. When the calculated similarity score is compared to the verification threshold, Mr. Gonzalez is denied access. Since Mr. Gonzalez really is Mr. Gonzalez and is not an imposter, the biometric system has erred and produced a false rejection. As Mr. Gonzalez and his business colleagues continue to use the biometric system, we would see over time how often anyone received a false rejection.

In Figure 11.12, we see someone else claim to be Jose Gonzalez. In this case, the imposter submits his vein pattern identifier for evaluation, and the system calculates a similarity score of 62, well below the threshold of 95. and denies him access. This is exactly what the system is supposed to do. However, over time multiple imposters claim Mr. Gonzalez's identity numerous times, and eventually one of them is successful and gains entry.

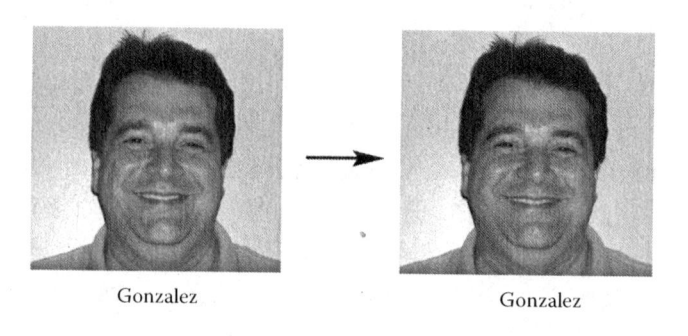

Gonzalez → Gonzalez

Figure 11.11   1:1 match.

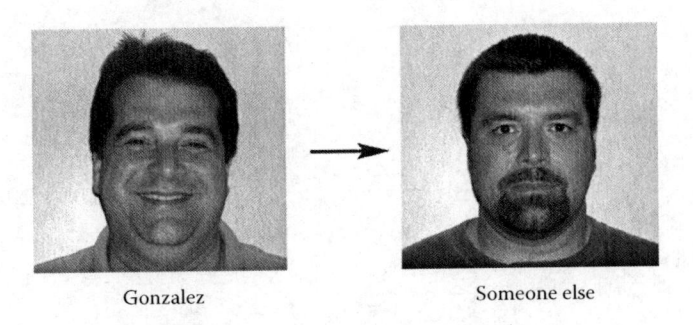

Figure 11.12 1:1 nonmatch.

The biometric system scores him a 95.5 and agrees that the imposter is Mr. Gonzalez—but, of course, he is not.

This is called a false acceptance. And if we gather information regarding how often the biometric system had approved a false claim, we could calculate a false acceptance rate (FAR), which measures the percentage of times the biometric system accepts an imposter. A false acceptance happens when a sample template from an unauthorized individual is incorrectly matched to a reference template of an authorized individual. This rate can be restated as a probability that a biometric system will incorrectly verify an individual or will fail to reject an imposter. For a verification system, it can be estimated as

(the number of false acceptances)/(the number of imposter verification attempts).

Authentication accuracy is an important determinant for biometric systems. Defining and achieving accuracy is a prerequisite for assuring security of the biometric system. Biometric system designers can accommodate a desired level of accuracy by setting the verification threshold to whatever is deemed an appropriate level. Most biometric systems have the capability of making threshold adjustments in relation to their desired level of sensitivity.

The sensitivity of the system might be set so high as to require nearly perfect matches of reference and live data. This most probably would reduce the FAR significantly, but it would most likely increase the FRR considerably as well because the FAR and FRR are mathematically connected. On the other hand, lowering the verification threshold such that the reference and sample template only approximate each other would result in fewer false rejections but potentially higher false acceptances. A biometric system can only operate on one threshold setting at a time. Therefore, adjustments for false rejections or false acceptances are generally made with respect to the application supported and the security level needed.

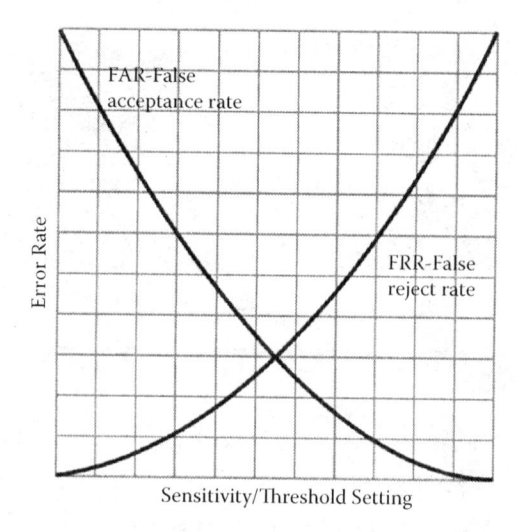

Figure 11.13   Error trade-off graph.

If we were to plot the FAR and the FRR graphically, as is done in Figure 11.13, then we would create an error trade-off graph. This graph is used by biometric system operators to help determine where to set their threshold sensitivities. The optimal threshold setting depends on what the biometric system operator is trying to achieve in terms of error trade-offs. With real-world applications, the FAR and FRR are traded against each other by manipulating one or two parameters, as calculated in an equal error rate (EER).

Figure 11.14 illustrates a DET (detection error trade-off) curve that depicts how, in biometric systems such as one that is public facing, the FRR rate might be set very low so as not to irritate legitimate users, while accepting the risk that an unauthorized person might be allowed access. FAR is to security what FRR is to convenience. For other applications, such as a high-security facility, the FAR might be set very low to ensure against potential intruders while tolerating the employee hassles associated with a higher FRR. In these latter situations, human intervention may be needed to remedy the false rejects.

Attacks on Biometric Systems and Their Remedies

As long as people create security systems to safeguard information and property, there will always be individuals or organizations who will attempt to circumvent that security. An imposter who tries to access a system for which he is not authorized or an individual who attempts to enroll in a

Figure 11.14 DET curve.

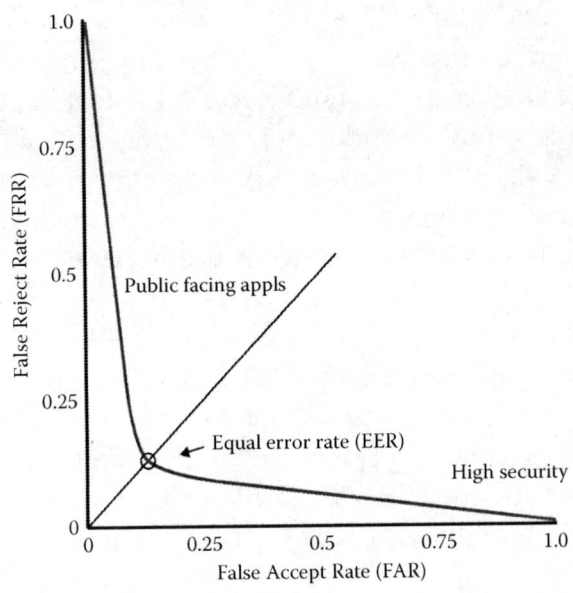

Note: Equal error rate is the point at which FAR equals FRR

biometric system more than once, creating multiple identities, are examples of the types of fraud that a biometric system must thwart. There are multiple ways to attack biometric systems depending on the modality and the quality of the overall system.

Attacks on biometric systems include spoofing,[20] replay attacks, and template database attacks. Some biometric systems have been defeated by submission of fake hands and fingers (with lifted fingerprints) or facial photography. Artificial fingers or other fake images can fool some biometric sensors in order to gain entry. Cryptographer and Professor Tsutomo Matsumoto of Yokohama National University gained international notoriety demonstrating how relatively easy it was to spoof *some* fingerprint systems using "gummy fingers,"[21] which are created using gummy bears as the gelatin base. Spoofing requires some basic knowledge of the targeted biometric modality, and sometimes it needs some level of collusion either with enrollment operators or with the targeted "victim" (referred to as an insider attack).

Antispoofing measures include use of accompanying passwords/PINs, smart cards, supervised enrollment,[22] and enrolling several samples. These are the most obvious low-cost and high-return antispoofing solutions. In addition, one might employ soft biometrics; liveness detection, including challenge/response solutions; or multibiometric systems.

By their nature, biometric devices are self-service appliances. So, an unattended biometric device might be vulnerable to a physical attack if there were a way for the attacker to open the device to defeat it. For this reason, many biometric readers now include tamper-resistant countermeasures such as special locking devices and/or alarms. Antitampering solutions are among the simplest and most cost-effective ways to thwart some common replay attacks. Other remedies include challenge/response systems, encryption/stenography, and time stamping.

Those systems that have networked biometric sensors that transmit to various processors might be intercepted. That is why all biometric data transmissions must be encrypted. In addition to thwarting efforts to fool a given biometric, there must be safeguards in the system itself to guard against playback attacks and other electronic attack techniques.

Yet another method to undermine a biometric system is to attack the other components of the overall security system. Security is only as strong as its weakest link, and the biometric component can be rendered ineffective if fraudsters can work around the biometric system. A good example of this is the high number of incidents of "tailgating" that occur at some enterprises that use biometric access. That is, someone who has gained entry to a building or office by authenticating himself holds the door open for others who bypass the biometric reader. A corollary to this is multiple incidences of exception handling, which may introduce a security weakness as procedures to work around the biometric system become institutionalized in the enterprise over time.

Detecting and addressing biometric fraud are challenging, but thwarting attacks on biometric systems is critical to maintain the confidence of the system stakeholders. Generally speaking, the vulnerabilities of biometric systems are offset by a variety of remedies. For the most part, the biometric industry has anticipated these threats, and most vendors have contrived and tested solutions to counter them successfully. It is somewhat analogous to the security software industry, which works tirelessly to provide software solutions that safeguard computer files. Just as computer viruses sometimes succeed in penetrating computer files, attacks on biometric systems are sometimes successful. However, that success is usually short-lived as biometric firms offer new, improved countermeasures.

It is important to note that examples of successful attacks on biometric systems, no matter how sensational, do not discredit the value of biometrics any more than the potential for a thief to steal one's house or car keys neutralizes the value of a locked home or car. People will continue to lock their possessions to safeguard (somewhat) against the threat of unauthorized entry.

Traditionally, data security has relied on something you know (user name and password) to grant information access. For at least the past four decades, physical security has relied on what you have (e.g., badges, keys, or access cards). These solutions have proven relatively ineffective, as they are easily lost, replicated, or stolen. Because of these limitations, solutions that rely on "who you are" easily trump the "what you have" solutions.

Summary

Biometrics can provide the connectivity between a person and his identity by linking to his measurable characteristics. Authenticating a person's identity within an acceptable probability is the primary value of biometrics to a physical security system. Biometric technologies are rapidly finding acceptance as the foundation for highly secure physical access solutions. The use of biometric technology in physical security applications by government, corporate enterprises, and individual citizens will continue to grow and will do so at an accelerating pace for the next few decades. And as long as the biometric systems deployed have privacy designed into their data collection, storage, and retention, and it is accomplished openly with individual consent, then biometrics can remain privacy enhancing. Additionally, when biometric solutions are properly implemented, they can usually provide greater physical security and convenience than alternative technologies.

Notes

1. Strong authentication refers to two-factor or three-factor authentication to deliver a higher level of authentication assurance. Authentication and verification are interchangeable terms.
2. Nonrepudiation is the assurance that a party in a transaction or similar event cannot refute or deny the validity of that event.
3. This assumes a noncompromised biometric system.
4. Spoofing is the process of defeating a biometric system through the introduction of fake biometric samples. This is a somewhat narrower definition than some biometric writers employ.

5. Capacitive sensors form fingerprint images on the dermal layer of skin by using principles of capacitance (generating a small electric current); there are passive and active capacitance sensors.
6. Ultrasound sensors use medical ultrasonography to create images via high-frequency sound waves that penetrate the epidermal layer of skin.
7. Optical sensors refer to the use of visible light to capture a fingerprint pattern image; this technique includes the frustrated total internal reflection (FTIR) method, which is the oldest and still the most used.
8. MSI, or multispectral illumination, is an optical method that looks at the skin's surface and subsurface features, capturing raw images by employing illuminating lights of different wavelengths and different polarization conditions.
9. Three-dimensional contactless is a finger imaging method that uses remote sensing to capture ridge-valley patterns. There is no physical contact between the finger and the sensor. While the technology continues to advance, it remains somewhat vulnerable to high-quality spoofing (e.g., using a fake gelatin-based finger with someone's lifted fingerprint).
10. The 3D ultrasound is based on matching paired images using internal fingerprint structures; it is difficult to spoof and has more tolerance to external conditions than most methods.
11. See www.eds.com/industries/homeland/downloads/idmanagement.pdf (accessed July 2009).
12. The INS Passenger Accelerated Service System (INSPASS) card uses hand geometry for identity verification purposes. The INSPASS program facilitates airport congestion by speeding up the Customs process.
13. The term *chaotic* refers to the dependence on the conditions in embryonic genetic expression. Morphogenesis is the process of cell development into different tissues, organs, or structures.
14. Read distance is the distance from the biometric sensor to the subject's physical attribute.
15. Some people refer to vein pattern recognition as vein geometry.
16. Technically, retina scans are a form of vascular pattern recognition since they use the pattern of blood vessels in the back of the eye to identify individuals; the same is true of facial thermography. For this reason, the term *vein pattern recognition* seems a more apt description.
17. There is no hard definition that is universally accepted; however, for vein pattern use, light that falls into the wavelength of approximately 700–1200 nm is usually classified as near-infrared light.
18. At IBG testing, the Hitachi finger vein and the Fujitsu palm vein both scored an availability rate of 99.92% based on enrollments.
19. Similarity score is a value returned by a biometric algorithm that indicates the degree of similarity or correlation between a biometric sample and a biometric reference.
20. Synthetic fingerprints can be created on the surface of a variety of materials, including gelatin, latex, and silicon.

21. Tsutomo Matsumoto's paper, "The Impact of Artificial 'Gummy' Fingers on Fingerprint Systems" (January 2002), became famous as the inventive professor traveled to international conferences and demonstrated how he could lift a fingerprint from a pane of glass, overlay it on a gelatin substance resembling a finger (as found in gummy bears) using an electron microscope, an inkjet printer, and Photoshop software.
22. User supervision is highly effective in ensuring uncompromised enrollments, but biometrics' strength lies in its self-service automation (except during enrollment).

Chapter 12

Security Guard Force

Paul R. Baker and Daniel Benny

Establishing a Security Guard Force

When establishing a security guard force, the security professional must first make a determination with regard to the size of the security guard force that will be required. The need for a security guard force must be established and will be based on several factors. These factors include a physical security survey of the facility to be protected by the security guard force, the duties and functions of the security guard force at the facility, size and mission of the facility, hours of operation, number of employees and visitors, security threat to the facility, and the budget.

The physical security survey and the physical security measures to be utilized will have an impact on the number of security officers that will be required to provide adequate protection of the property. The use of intrusion detection systems, security cameras, fire protection systems, and access control such as proximity card readers may reduce the number of security officers required to patrol a property. If there are no or limited physical security measures, there will be a requirement to establish a larger security guard force in order to secure the facility effectively. The use of more physical security measures may allow for the reduction of the size of the force. Regardless of the level of physical security protection, there will in almost all cases be a need for security officers to monitor intrusion, fire, access control, and camera systems. There may also be a requirement for security officers to be able to respond to the various alarms or activity observed on security cameras.

Mission and Duties of the Security Guard Force

In determining the size of the security guard force, the mission and duties of the security guard force must be determined. The primary duty of a security guard force is to provide proactive patrols of the property in order to prevent losses, respond to emergencies, provide assistance to staff and visitors, and enforce company regulations. These patrols may be conducted by numerous methods, including foot patrol and the use of vehicles such as automobiles, all-wheel drive vehicles, bicycles, Segways, or other special-use modes of transportation. This will depend on the terrain, weather, and other geographical features.

The security guard force may also be utilized to control access to the property. The access control may begin at the perimeter of the property at vehicle entrances. The security guard force would be responsible for obtaining identification of drivers and may also conduct inspections of vehicles entering the facility. Access control points required to be covered by security officers may also include visitor entrances, employee entrances, or delivery areas. There may also be high-risk areas inside the structure, such as a cash office, restricted areas, or the office area of the chief operating officer, that require the use of security officers to control access, based on the nature of the organization.

Escorts are often provided by the security guard force. These escorts could be for the transportation of money, high-value items, or company confidential information related to the business. These escorts may take place on the organization's property or off the property in the case of a money escort to a banking facility. Security could also provide escorts for visitors to the organization. This is especially true if there is a need to protect a company's sensitive areas and information. Providing security escorts to employee parking areas for employees leaving work during hours of darkness may also be a service that is provided by the security guard force.

Inspections of the facility for security threats, safety, and loss hazards are a function that should be performed by the security officers while on patrol. Depending on the size of the organization's property and the number of buildings, this may be a duty that would be performed by security officers on patrol, or additional officers may need to be hired to perform this important inspection function to ensure the security and safety of the employees and visitors on the property.

Investigations of losses, safety issues, accidents, violations of company regulations, and employee misconduct will require the attention of one or more full-time investigators if there is a significant case load based on the size and population of the work site. Investigations are a key function of the security department in the protection of the organization's staff and assets. This function needs to be considered when staffing the security department.

Protective service may be a full-time requirement for executives of the organization based on the profile of the organization, the executives, and the threat assessments. Full-time protective service coverage would include protection of the executive at home, when traveling, and at the workplace. Based on the situation, protection may not be full time but rather during travel or periods of high threat. In any event, the protection of key management needs to be considers when developing the security guard force.

Special events must also be considered in determining the size of the security guard force. If the organization has numerous special high-profile events during the year, there will be a need for additional security during those periods of time. It is important to ensure that there is security coverage to meet the requirements of the day-to-day security duties in addition to those added requirements during special events.

Monitoring of intrusion detection and fire safety systems, security cameras, and access control is an import function of security. The establishment of a proprietary security communications and monitoring center (Figure 12.1) to dispatch security staff, answer security-related calls, and monitor the security, fire safety, cameras, and access control points will require the hiring of additional security officers to perform this vital function. These positions should be staffed by trained security officers who can be rotated between patrol functions and monitoring duties. This is critical since a

Figure 12.1 A security communications and monitoring center.

trained security officer will be more effective at responding to security calls and situations arising while monitoring the security and safety systems than a person hired only to work in the proprietary communications center.

It is also important not to have an individual monitor such systems for more than 2 hours. A security officer will become less effective at monitoring a security camera if the assignment lasts more than 2 hours. By having security officers working in the communications center, they can be rotated to the patrol function after 2 hours in the communications center.

The final function to consider is the administrative duties associated with a security department. These duties will include securing security department records, processing of internal violations such as parking tickets, and preparing correspondence, monthly reports, and any other administrative duties that may be required. These positions may be in the role of secretary to the director and administrative clerks.

Based on a review of all the possible duties and functions of a security department that have been described, a final determination of what services and duties the security department will perform needs to be made. The functions and services the security department will perform will have an impact on the size of the security department.

Profile of the Facility to Be Protected

A review of the profile of the facility that needs to be protected is necessary when determining the size of the security guard force. The mission of the facility, its size, hours of operation, number of employees, visitors, and security threats are key elements that must be considered to make a determination as to the size of the security guard force.

Mission of Facility

The business's mission, type of business, and activity that is conducted at the facility to be protected will have an impact on the risk to the facility and staff. There may be risks to staff, customers, or visitors. The risks could also be to property, products, or the operation of the organization. The risks will, of course, vary based on the mission. If the facility houses high-cost products, there may be a high risk of theft or burglary. If the facility is an abortion clinic, there could be risk to the doctors, staff, and property by terror-related antiabortion groups. If the facility is an airport, there would be the risk of hijackings or other terrorist activity targeting the aviation community. The activity of the facility must be examined when determining the type of physical security and security guard force staffing.

Security Threat

The security threat to a facility will be based on numerous factors, including the mission of the facility, its size, hours of operation, and the location of the facility. The local crime rate and previous crime and losses against the organization must also be evaluated to determine the current risk to people, the facility, and organization.

Size of Facility

The size of the facility, including the square footage of buildings, the number of floors in the buildings, the total number of buildings, and acreage of the property to be protected must be calculated in determining the number of security officers required to provide adequate protection. Based only on size, one small, single-story building on a half acre of property will require less security coverage than a college campus with 20 buildings on 20 acres of property. Such considerations need to be taken into account when planning security officer protection.

Hours of Operation

Hours of operation will impact security guard force coverage. If the facility is only open 8 hours a day and is then secured with an intrusion detection system, it obliviously will require less security guard force coverage than a 24-hour operation. A hospital or university campus that is occupied 24 hours a day will require more security officer coverage than a retail store operating 12 hours a day. As hours of operation lessen or expand based on the specific situation, the level of security coverage will also need to be adjusted to meet the need of the organization.

Number of Employees/Visitors

The number of employees located at a facility can have an impact on the size of the security guard force depending on the services offered to the staff. If the workforce is large and security will be offering escorts on a campus or to employee parking areas, then there may be a need for more security officers. If a large number of employees need to be screened entering the facility by use of metal detectors and the checking of packages, then more security needs to be scheduled than for a facility with less staff and no screening requirements.

Proprietary Security Guard Force

A proprietary security guard force is one in which the security officers are employees of the organization. A proprietary security guard force may be full time, part time, or a combination of full- or part-time positions. Based on the type of position, officers may qualify for full or limited company benefits such as medical coverage, insurance, vacation, and sick leave.

The advantages of proprietary security include the quality of personnel, degree of control over the security program, employee loyalty to the company, and prestige for both the employee and the company. A proprietary security guard force would be provided better training, which would be more specific to the operation, and better training equates to better performance. There would be a sense of employee loyalty, which will create a stronger sense of ownership. Officers see themselves as a part of the team and are willing to go the extra mile for the benefit of the company and other employees. They see themselves with a vested interest in the long-term success of the company. Proprietary guards benefit from esprit de corps and a sense of community. Since it is possible to pay better wages to proprietary employees, turnover might be lower in proprietary guard organizations. Proprietary guards have a knowledge base concerning the operation from being embedded into the corporate culture. The knowledge that comes with years of experience is another advantage of the propriety guard force

Utilizing proprietary security allows companies to design, field, and manage a security program to its particular needs. "Many managers feel that they have a much greater degree of control over personnel when they are directly on the firm's payroll."[1]

Employee loyalty is much greater in a proprietary system. Contract security personnel are often moved around between different clients, making it hard for them to develop relationships and a sense of loyalty to a particular client. In the end, it is the contract company, which "butters their bread," whom they protect—not the client.

Proprietary security personnel enjoy a greater degree of prestige than is afforded their contract brethren. Many contract personnel are derided as "rent-a-cops" and are neither as highly trained nor as professional as proprietary personnel are. In addition, "many managers simply prefer to have their own people on the job. They feel that the firm gains prestige by building its own security guard force...rather than renting one from the outside."[2]

However, proprietary security operations have their disadvantages, including: cost, administration, staffing, impartiality, and expertise.

Establishing a proprietary service in comparison to contacting a contract guard service starts with the time it would take a firm to establish its own program—from selecting a head of security, filling of numerous positions, and the specific training that would be needed. Businesses would spend a considerably larger amount of money and time on in-house services where they would have to offer competitive wages and benefits to compete with other businesses and security firms. Another disadvantage is the possibility that more permanent in-house guards have the potential to form friendships and become somewhat partial to other employees, which could result in favoritism and a lack in their performance of duty.

Contract security costs less than proprietary security because, typically, proprietary personnel earn more than contract personnel, due to the prevailing wage at their company. Contract personnel generally have fewer benefits than proprietary personnel do. Start-up costs for proprietary systems also make contract security services cheaper.

The administration required to operate a proprietary security operation includes recruiting, screening, and training security personnel, as well as maintaining logs, audits, and other security program components. "There is little question that the administrative workload is substantially decreased when a contract service is employed."[3]

The disadvantages of a proprietary security guard force are that it takes longer to hire staff and costs more as the organization must place advertisements for recruitment, conduct pre-employment background investigations, and supply uniforms and equipment. There is also the cost of a complete benefits package.

Another disadvantage is that once the security officer makes it past the probationary period, it is more difficult to terminate him or her. To do so, all actions must be documented and progressive disciplinary action must be utilized unless the offenses are serious enough to warrant immediate termination.

Contract Security Guard Force

A contract security guard force is one that is made up of security officers working as employees of a licensed security or investigative firm that provides security service on a contract basis and who are not on the payroll of the organization utilizing their service. In most states, contract security providers must be licensed, so it is important to select a firm that meets this legal requirement.

Contract security can adapt staffing levels more easily than a proprietary system, as the need for staff rises and falls due to sudden or unexpected circumstances. The cost of hiring, training, and equipping proprietary staff makes rapidly increasing or decreasing staff levels more cumbersome than if contract services were used. With contract guard services, the cost for salary, insurance, administrative costs, uniforms, and benefits are all rolled up in one hourly price. This is helpful for budgeting purposes and there are no hidden costs.

Contract security employees are seen as impartial because they are paid by an outside company to enforce a set of policies and are not as likely to deviate from procedures due to personal relationships or pressure. Guard companies do not have the loyalty of proprietary officers and in this regard they have no problems with enforcing rules and regulations. There is no bond between the employees and the guard company and if there is found to be a buddy-buddy system starting to develop, a phone call to the account manager can replace the guard easily.

Contract security companies focus on security as a business, not a supporting function of business. This gives them an advantage over proprietary operations. "When clients hire a security service, they also hire the management of that service to guide them in their overall security program."[4] It may take time for a company to develop a proprietary security organization with the wide-ranging level of experience that comes with a contract security company. According to Alan Stein, vice president of Allied Barton, contract guard service companies offer substantial advantages when it comes to training. "A contract company will usually have a dedicated training department," he says. "The department is responsible for developing general and specialized training materials."

When it comes to staffing needs, guard companies already have a process established for the recruitment and hiring of staff. Their business is to have bodies available for assignment. This relieves the company of paying overtime when proprietary employees leave or take vacation. With a contract guard service, when you need three guards, you ask for three guards. When you do not need three anymore, you ask for two. Another big benefit is that if a contract officer does not work out, he or she can easily be replaced. You just call your account manager and ask for a replacement.

Guard companies take care of liability issues by having their guards licensed and having the company bonded. Most states require that security guards be licensed. Requirements vary widely but in most states, applicants must be at least 18 years old, pass a background check and drug test, and complete classroom training in such subjects as property rights, emergency procedures, and detention of suspected criminals.

The disadvantage of a contract guard service is the quality of the guard you would receive. They are the first line of protection and if they do not have the training or knowledge of your operation, then it is nothing more than having a mannequin sitting at a post. They are not highly paid and they may be temporary; there is no continuity to their service and no ownership of the post and operation. In other words: you get what you pay for.

Hybrid Force

Sometimes contract security officers make sense. Sometimes, proprietary guards make sense. Sometimes, mixing both together at the same institution makes better sense. The hybrid system allows an organization to maintain more control over its security program, while achieving the cost savings and administration reduction associated with contract security. The use of hybrid systems is considered a workable solution, as it affords the benefits of both contract and proprietary security, while mitigating the downsides of both. It allows easy access to contract guards when the need arises, bringing in personnel that can work in conjunction with proprietary staff.

In these situations the contract staff is using the position as a jumping pad to be employed by the organization. This allows the company to review the services of a contractor and when requirements for additional proprietary staff or replacement of proprietary staff come up, the security manager has already witnessed the capabilities of potential new hires and will make it an easier fit into the organizations.

Security Guard Uniforms

Traditionally, security officers wear a uniform. A uniform is a symbol of authority and allows the security office to be identified easily during an emergency or when assistance is required by staff or visitors to an organization (Figure 12.2). The most common security uniform is slacks and a short- or long-sleeve police/military style shirt with a security patch, name tag, and a badge where authorized by state or local laws. Utility belts are often worn to carry security and protective equipment such as keys, radios, flashlight, OC spray, baton, or firearms. During colder weather, there are a variety of light- and heavyweight water-resistant jackets and coats that can be utilized. Patches, name tags, and badges are also placed on the outer garment for ease of identification. Headwear is also part of the security uniform and can be a more formal eight-point cap, trooper hat, or ball cap style with badge or security insignia placed on the front of the headwear.

Figure 12.2 Security must be easily identifiable to employees and visitors alike.

Special uniforms may be utilized for special assignments or weather conditions. This may include shorts for bike patrol in warm weather conditions or tactical uniforms for special operations or in remote areas. In some organizations, security officers may be dressed in business attire with a badge or name tag placed on the front of the jacket. For protective service or undercover operations, the security officer will dress to conform to the environment. This may include formal dress, business dress, business casual dress, or street clothing.

The appearance of the security officer, especially when in uniform, is critical in presenting a professional and authoritative image.

Security Guard Identification

Just as the security uniform provides a symbol, so does security guard identification. Badges, where authorized by state and local law, are a universally recognized symbol of authority. Shoulder patches also add to the authority of the security officer and identify the organization with whom he or she is employed. The most important aspect of security identification is a photo identification card to be worn on the uniform or carried in a case to provide for positive identification of the security officer and the organization for which he or she works.

This professional image begins by wearing the assigned uniform in a proper manner. Security officers should only wear the uniform items issued and should not be permitted to customize by adding or deleting aspects of the issued uniform. If this takes place, the security officers are no longer "uniformed" and it is unprofessional. The security uniform must also be clean and pressed at all times when the security officer reports for duty.

There should be grooming standards for a security officer who is wearing a uniform. These grooming standards should include hair length and style, facial hair such as beards, and the wearing of jewelry, earrings, and other piercings visible when wearing the security uniform. Proper hygiene should also be addressed in the standards.

The demeanor of the security officer in uniform is also important. Exhibiting good posture and a professional attitude will project a professional image.

Security Guard as an Authority Figure

Security uniforms and identification allow the security officer to be identified as an authority figure, but uniforms and identification alone do not provide the security officer with such authority. The authority must come from legal codes that apply to security officers depending on the state in which they operate. The authority also comes from the organization for whom they are employed with regard to enforcing company regulations on the organization's property and the ability of the security officers to enforce such regulations. This legitimacy must also be based on the proper use of such authority.

Security Guard Protective Equipment

Where authorized by law, protective equipment may be considered for the security guard force. The type of protective equipment utilized will be based on the threat level, location, and mission of the security guard force and may range from handcuffs to the carrying of firearms. Many states require specialized training before someone is authorized to carry various types of protective equipment.

In Pennsylvania, for example, security officers who carry a baton or firearm must complete what is known as the Lethal Weapons Act 235 Course. To attend the 40-hour course, the student must submit to a criminal background check and medical and psychological evaluations. The 40-hour course covers the legal aspects of carrying a weapon, the authority of a security office, use of force considerations, and the Pennsylvania Crimes Code. Students must pass a written test and qualify on the firing range to become certified under the Lethal Weapons Act. It is important to know the requirement with regard to carrying a weapon in the state in which security officers are operating to ensure compliance with the laws of the state.

Handcuffs are important should security guards be required to make a citizen's arrest in the performance of their duties. Handcuffs provide a means to secure an individual who becomes violent either before or after a citizen's arrest. The use of handcuffs in such situations provides for the safety of the security officer and the public. Handcuffs should be of good quality and have the capability of being doubled locked. The double-locking mechanism prevents the handcuff from being tightened on the suspect by accident or by the suspect to then claim an injury from their use.

Oleoresin capsicum, or OC spray, is a lachrymatory irritant agent that can be carried by security officers. It provides a nonlethal method of self-defense for the security officer and is very effective in most situations. Security officers should be certified by the manufacturer of the oleoresin capsicum product to ensure proper use and for liability purposes.

Batons have been carried by security for over a hundred years and they can be used as a defensive and offensive protective tool. When used offensively, they are considered a deadly weapon. Batons come in various styles to include the traditional striate baton, the collapsible ASP baton, and the PR-24 full-size or collapsible model. Certification should be obtained from the manufacturer of the particular baton that is carried for proper use and liability protection.

Firearms may be carried by the security guard force based on the legal requirement of the state and threat and mission of the security guard force at a particular location from which they are operating. A revolver or semiautomatic firearm may be carried and in some situations security officers may also carry a shotgun. In addition to state legal requirements for qualification and certification to carry a firearm, security officers should be trained and qualify with the weapons and ammunition they carry at least once a year. Many security departments require such training and qualification twice and up to four times a year.

Armed or Unarmed?

The question is what type of guard should be used: armed or unarmed? Armed guards should only be used in situations requiring protection of money, jewelry, art works, firearms, or other objects that might attract armed robbers. Armored car companies are employed to pick up and deliver large quantities of cash. Protect the money that is exposed to the public by making frequent safe-drops, keeping cash register drawers shut whenever possible, and enforcing rigid money-handling procedures. Other valuables should be physically secured, if possible, before exposing them to the public.

It is generally not a good idea for retail establishments to employ armed guards in public areas. If someone is intent on robbing a retail business, there is no way to tell who that person might be, and the guard with the gun becomes the first target. An unarmed guard will deter thieves just as well without becoming a source or cause of violence in the event of a holdup. What are you expecting an armed guard to do when confronted with a situation? When would you want him or her to fire his or her weapon? What liability will you and your company be placed under if a guard injures a criminal, bystander, employee, or another guard? In most cases, the use of an unarmed guard is the most prudent application of deterrent that will be needed.

Security Guard Training

One of the most important aspects in the management of a security guard force is to ensure that the security officers are effectively trained to meet any state regulatory requirements as well as security industry standards of training. Such training will promote professionalism within the security guard force and reduce the liability risk. Security guard force training can be accomplished by on-the-job experience and training and through the use of various formal education methods.

On-the-job experience and training comprise a structured and documented approach in instructing the new security officer with regard to day-to-day duties as a security officer. Each new security officer should be assigned to a mentor. The mentor may be a supervisor, lead officer, or training officer who will guide the new officer through daily activities, providing instruction on how to perform his or her duties. As each new task is learned, it should be documented in a written training record for each security officer.

As the security officer accumulates time in the profession and the various security assignments, he or she will gain knowledge and proficiency in the profession. Other on-the-job educational tools may include having the security officer take part in organizational meetings and committees to expand professional knowledge. This may include being part of the organization's safety committee or meetings related to special events that might be scheduled.

In addition to on-the-job training, more formal educational methods should also be applied. This may include company assistance for the security officer to obtain a college degree in security or criminal justice. In-service training can also be used, where the security officer is provided with information in a classroom environment covering security procedures, report writing, patrol methods, or court testimony. In-service training can also be used to provide the security officer with various certifications, such as first-aid and CPR, handcuff, OC, or baton certification.

Another option for education is to have the security officer take part in self-study by online proprietary training or a website offering free training, such as Homeland Security FEMA Academy. Time for such online training can be provided during the work schedule or it can be accomplished off duty. Directed reading is another source of education; articles or documents related to security are made available to the security officer, who is required to sign off that the document has been read.

In order to ensure that the security guard force is professionally trained, a security training program needs to be established and mandatory training needs to be provided to all security officers. All state regulatory training requirements, where applicable, must be completed. It is important that all training completed by each security officer be documented in the security officer's training file. This will allow for the tracking of the training to ensure that it has been completed; such documentation is required by regulatory agencies or can be related to liability issues.

Professional Security Certifications

Professional security certification can be obtained and is of value to those in the security profession. The American Society for Industrial Security (ASIS International) has developed several professional security certifications that are accepted by United States Homeland Security, as well as nationally and internationally.

The certified protection professional (CPP) has been established for individuals working in security supervision and management. Upon successful completion of the comprehensive examination covering all aspects of the security management profession (such as management methods, security guard force management, legal issues, investigations, physical security, protective service, terrorism, and budgeting), the designation of CPP is bestowed.

The professional certified investigator (PCI) was established for the security officer or private investigators. Upon successful completion of the examination, which covers all aspects of security and private investigation to include investigative methods, legal consideration, and interview methods, the designation of PCI is bestowed.

The physical security professional (PSP) designation is designed for those in security who have responsibility for physical security within their organization. The examination covers intrusion detection systems, barriers, security cameras, and lock and access control. Upon successful completion of the examination, the designation of PSP is bestowed.[4]

Personnel Issues

The first step in the recruitment of a security guard force is to establish a position description identifying the required duties and responsibilities, as discussed earlier in this chapter. Experience, education, and physical ability requirements also need to be established—not only to ensure that the best individuals for the positions are hired, but also to comply with the Americans with Disabilities Act. Recruitment of security officers should come from outside the organization rather than from current employees in other departments. Hiring from within the organization can create conflict of interest with co-workers and departments that they worked with in the past.

The background investigation of the final candidates is critical to ensure that individuals meet all the requirements of the position and that they are qualified for the position, ethical, and trustworthy. Previous employment, education, criminal history, credit reports, driving history, military history, and references should all be part of the background investigation.

When making the final selection, also investigate the person's motives for applying for the position. Examine his or her ability to deal effectively with other employees and visitors to the organization as it relates to security issues.

Security officers employed by the organization need direction as to their duties and responsibilities. This is accomplished through the development of a *Security Policy and Procedures Manual.* This document provides the security officer with detailed guidelines with regard to responsibility and duties within the security department. It also gives him or her the authority to perform such duties on behalf of the organization.

The *Security Policy and Procedures Manual* should be written in a brief, clear, and concise manner so that it can easily be understood. A written copy should be given to each security officer. The security officer should be required to sign for the receipt of the manual and then be tested on it. This will document not only that the officer received a copy, but also that he or she has read and understands the contents of the publication. This is vital with regard to liability issues and the discipline of security officers who fail to perform their duties in a proper manner.

The *Security Policy and Procedures Manual* needs to reviewed and updated at least once a year or sooner when needed. This would include the changing of any security policies and procedures.

Always keep the manual current and never put anything in the security policies and procedures that cannot be accomplished by the department or security officers. Should there be civil action, the organization will be held to its own standards established in this document. That is why it is important that what is written in the manual can be and is accomplished.

Within the *Security Policies and Procedures Manual,* a "Standard of Conduct" for the security officers needs to be established. This will be the basis of evaluating the security officers' conduct on the job. The goal is to inform the security officers of what is expected of them and the penalties for not adhering to the code of conduct.

Performance and conduct of the security officer need to be evaluated and dealt with through progressive disciplinary action. Some conduct offenses, such as theft, falsification of reports, and violent behavior, may require termination. However, for most conduct offenses and performance issues the correction of the problem is the best solution.

There are five steps in progressive disciplinary action that allow the organization to take corrective action and allow the security officer the opportunity to improve conduct and/or performance:

- *Counseling.* This is a process of advising the security officer of a problem with performance or conduct and discussing with him or her what is required to correct the problem as well as determining why the failure occurred. A review of the policies and procedures or retraining is a common solution to correct the security officer's performance or conduct.
- *Oral reprimand.* If a security officer has a second offense with regard to conduct or performance, the next step is an oral reprimand. The organization has moved beyond the counseling and education phase and has placed the security officer on notice that he or she must correct the issues at hand.
- *Written reprimand.* For the third offense, the security officer's actions are formally documented in written form advising the security officer of the seriousness of the situation. The security officer is to be advised that if he or she fails to correct performance or conduct, a suspension or termination from the position could take place.
- *Suspension.* The suspension step is a serious action in which the security officer will lose pay and will be given one last chance to correct his or her performance and/or conduct.
- *Termination.* Termination is the final step after all other steps have failed to correct the security officer's performance or conduct and he or she must now be removed from the position. When terminating

the security officer, do not allow the officer to remain on the job site. Collect issued identification, keys, equipment, and uniforms and escort him or her off the property.

Progressive disciplinary action is an effective means of correcting issues with security officers and gives them the opportunity to improve and continue to be a part of the security guard force. It is also an excellent management tool to remove a security officer from the security guard force when his or her performance or conduct does not meet the department's standards.

Summary

When considering the burden to a company's security cost, guards are added expenses. Expenses that must be factored include wages, training, sick call, vacation, liability, etc. Each expense factors into the overall security picture. If a company is not willing to accept and address these expenses, the chances of an adversary bypassing detection devices increase. The guard is the human factor that is both good and bad, regardless of what type of guard force the company provides (proprietary, contract, or hybrid).

On the good side, the guard has the ability to think and react on the fly upon assessment of an alarm. On the bad side, human nature allows for such activities as sleeping, daydreaming, and corruption. Overall, using guards at entry control points increases the chances of detecting adversaries attempting to gain access. When used with electronic access control and surveillance systems, the two complement each other to create a strong physical protection system.

References

1. Fischer, R., E. Halibozek, and G. Green. 2008. *Introduction to security*, 53. New York: Elsevier.
2. Ibid., 54.
3. Ibid., 52.
4. Ibid., 53.

Chapter 13

Central Station Design

Paul R. Baker

Developing an Operation

After all the cameras, alarms systems, and locks are installed throughout your facility, the question remains: Who will monitor and respond to the alarms? As a security manager, you have a choice between an on-site monitoring station or a third-party off-site monitoring service.

If an organization goes with an in-house monitoring setup, then it will have the central station located on its property and monitored by a proprietary or contract guard service. The central station will have all alarms, CCTV, and access control signals tied directly into the operation center.

The central center, also known as the security console center, security control center, or dispatch center, is an area that serves as a central monitoring and assessment space for access control, CCTV, and intrusion detection systems (Figure 13.1). In this space, operators assess alarm conditions and determine the appropriate response, which may entail dispatching of security forces. Normally, the central station is staffed by trained personnel 24 hours a day, 7 days a week.

Maintaining a 24/7 security control center requires at the minimum two officers per shift. They are responsible for monitoring alarms, access control, CCTV, and fire. They will be required to dispatch officers to investigate alarms, disturbances, and unknown events.

In this circumstance, a hybrid scenario is a practical application in the use of proprietary officers within the security control center and contract

Figure 13.1 The central center, also known as the security console center, security control center, or dispatch center, is an area that serves as a central monitoring and assessment space for access control, CCTV, and intrusion detection systems.

officers out on the floor. Inside the control center, you need a higher level of commitment to the company and performance continuity in knowledge and action. The contract officers can be cross trained for times when there is a staffing issue dealing with the control center proprietary staff. This would be a support element that could be counted on for vacation, sickness, or temporary duties. And even though contractor management does not like being an employment warehouse, the contract officer can be viewed as a potential replacement in the event that a proprietary position becomes available.

The federal government utilizes the UL 1981 standard when dealing with high-security facilities for designating staffing levels at a central station:

> UL 1981 requires monitoring facility staffing be such that all alarm signals be acknowledged and the appropriate dispatch or verification action be initiated not more than 45 seconds after the monitoring facility receiver acknowledges to the alarm panel at the protected site that the alarm signal has been received.
>
> a. Staffing at the central monitoring facility shall be such that it is in compliance with UL 1981.

b. Staffing at the central monitoring facility shall exceed requirements in UL 1981. Staffing shall be such that all alarm signals be acknowledged and the appropriate dispatch or verification action be initiated not more than 30 seconds after the monitoring facility receiver acknowledges to the alarm panel at the protected site that the alarm signal has been received.[1]

The Police Scientific Organization has done research on the number of monitors that one control center operator can handle effectively. It indicates that no more than four to five monitors per person should be allocated. This is supported by its research on watching observers viewing one, four, six, and nine monitors, which showed accuracy detection scores of 85%, 74%, 58%, and 53%, respectively. Clearly, detection rates go down as people have to cover more information.

"Application and Design of CCTV"[2] identifies a "time and fatigue" factor. A person with an average IQ can only watch an average of 30 to 45 minutes of continued video imagery before he or she loses comprehension about what he or she is looking at. Also, the same person can watch up to a maximum of four scenes simultaneously without any comprehension at all.

The American Society for Industrial Security (ASIS) states that a single control center monitoring station should consist of no more than 15 monitors.

Design Requirements

If the facility is entirely under your control, the control center should be located on the main floor or in the basement of the facility, as long as the area is not below ground level and there is no chance of flooding. Entry will be controlled by an access control system and CCTV for viewing at the control center door. This will give the control center operators a view of who is attempting to gain access and only authorized personnel will be allowed inside the center.

One of the worst, though unfortunately most common, locations for a security operations center is in the reception area of the building it is protecting. During crisis situations, the foyer of a building becomes an incredibly chaotic place where most building occupants will be attempting to leave in a hurry, making it an unsuitable location to form an organized response, if only because of the noise involved. A good security operations center should be located centrally within a facility, away from active primary entrances and exits. A security control room should have heavy-duty reinforced doors that are to remain closed and locked at all times when not in use to keep the control room a safe haven in any situation.[3]

Mike Fickes, a security design specialist, recommends that a security operations center will need surveillance cameras and monitoring as well as recording equipment to review video feeds at a later date, particularly if they are ever needed in a legal proceeding or investigation. A solid suite of computer, telephone, and intercom communication systems is also highly recommended, as is comfortable seating for staff, since the control room should always have at least one monitoring security officer. Additional technology, such as remote door lock controls, can also be quite helpful and is frequently employed with magnetic door locks.

Ceiling height is also important. If you put six 52-inch monitors on the wall, you will need a ceiling at least 10 feet high. You should also provide enough space to set the workstations back at least 15 feet from the wall to give the operators a comfortable view.

Lighting should be dimmed, with lamps directed toward the ceiling. The reflected light reduces glare, which can make it hard to concentrate on large monitors and desktop computer screens.

Wired telephones and a phone desk large enough to accommodate a number of people have a place in the security center as well. During an emergency, communication with the outside world is essential. One or two of the phones should not go through the building's exchange but rather provide plain old telephone service (POTS) that works even when the building's power goes off and shuts down the building system.

If possible, place all digital video recorders and other computer equipment into a separate space where you can filter the air and control the temperature, which will extend the lifetime of the equipment.

Just like the building itself, the security center should have fire alarm and suppression systems, including sprinklers in the main room.[4]

The control center will be provided with primary and secondary sources of power. The secondary power should consist of a battery backup or UPS system that will provide for a normal load of at least 24 hours. It is also recommended that an engine drive generator system be used as a complement to the rechargeable battery system in order to keep uninterrupted power to the control center.

Typically, the only communication redundancy made is between the field panels and the system head-end. Redundancy between field panels and devices is cost prohibitive. A common method of achieving communication redundancy is achieved by running primary as well as backup RS-485 lines. If this is done, it is best to use different raceway routing schemes. Redundant communication paths are established in order to maintain communication through an alternate path if a component or link goes down.

Secondary Amenities for an Operations Control Center

Some secondary amenities for a security operations center include a full suite of handheld radio equipment, including coding chargers and emergency lighting systems, such as flashlights and lanterns. A well-stocked first-aid kit is also good to have on hand, preferably stocked with an automated defibrillator device for treating heart attack victims, provided that security personnel have relevant training. Emergency rations are also helpful to have on hand in the event of extreme disaster situations.[5]

Alarm Assessment

"Detection is the notification that a possible security event is occurring; assessment is the act of determining whether the event is an attack or a nuisance alarm."[6] This is where the central station makes its living. It has to make a decision to dispatch a response force to investigate the alarm or it determines that it is a nuisance alarm and documents and monitors further. If the central station has the ability to view the alarm with the availability of CCTV, it has an assessment capability. "An assessment refers to immediate image capture of a sensor detection zone at the time of an intrusion alarm," while "surveillance uses CCTV to continual[ly] monitor activity in a[n] area, without benefit of a[n] intrusion sensor to direct attention to the specific event."[7]

It takes a seasoned central station operator to know his or her surroundings and deal with the alarm components of the facility. Sometimes this can cause complacency if nuisance alarms are continual and not addressed. When the operator continually receives an alarm and clears or places the alarm in test mode, this can have serious repercussions if the alarm is

Figure 13.2 Alarm monitoring and a responsive guard force will prevent unwanted break-ins, theft, and potential loss.

actual. There is nothing worse then receiving an actual alarm and doing nothing. This is where a two-person requirement is necessary in a central station. With one alarm coming in, it requires both operators to clear the alarm. Redundancy sometimes makes the difference between complacence and compliance.

As depicted in Figure 13.2, the last thing anyone wants is for employees to come to work the next day to find their spaces broken into and then question why security did not do its job and respond to the alarm.

References

1. Unified Facilities Guide Specifications. 2008. Central monitoring facility staffing, 7. www.wbdg.org/ccb/DOD/UFGS/UFGS%2028%2020%2002.pdf
2. Pierce, C. 2002. Application and design of CCTV. www.lrc-inc.com
3. Mueller, D. The best practices for building a security operations center. http://www.ehow.com/list_7467979_practices-building-security-operations-center.html#ixzz1i1n9y5Xw
4. Fickes, M. 2011. How to design a security control center. http://www.buildings.com/ArticleDetails/tabid/3334/Default.aspx?ArticleID=12405#top

5. Mueller, Op. cit.
6. Garcia, M. L. 2001. *The design and evaluation of physical security protection systems,* 113. New York: Elsevier.
7. Ibid., 114.

Chapter 14

Government Security

Paul R. Baker

SCIF

In highly restricted work areas, where sensitive compartmented information (SCI) may be stored, there will be a requirement to construct a secured area. These storage areas may consist of government sensitive compartmented information facilities (SCIF) and special access programs (SAP) or closed areas where there will be a requirement to increase the security blanket to ensure stricter access to these areas. The physical security protection for a SCIF had been outlined in DCID 6/9, but has been updated to the Intelligence Community Directive (ICD) 705, which is the requirement for governmental physical security standards. ICD 705-1 is the established procedure for the construction and protection of government facilities for storing, processing, and discussing SCI, which requires extraordinary security safeguards and is intended to prevent any visual, acoustical, technical, and physical access by unauthorized persons.

Your organization may not be required to maintain government-classified information; however, the company's livelihood and your employment are tied to proprietary information that requires the same levels of security. A SCIF construction project should start with a construction security plan (CSP) to address the planning, design, and construction. All design and construction of the SCIF will be performed by US companies using US persons. During a renovation of an operational SCIF, construction personnel will need to be cleared or escorted by personnel cleared to the level of the SCIF.

Location

It is best to locate a SCIF in the center of an office floor where there will be no windows and the area has that extra layer of security. However, this may not always be practical. If the SCIF will be on a floor below the fourth floor of a building and windows will be part of the design, it can still be accredited, but there will be necessary construction requirements to meet the accreditation.

Design

Perimeter Walls

SCIF walls will consist of three layers of 5/8 inch drywall and will be from true floor to true ceiling. This consists of a 3½ inch stud with two sheets of drywall on the secure side and one sheet on the nonsecure side. There should be a minimum of R-19 insulation between the studs. Walls must be permanently constructed and drywall must be taped at all seams, mudded, and finished. The top and bottom of the drywall must be sound caulked on both sides of the dry wall (Figure 14.1a–d).

Doors

There will typically be only one SCIF entrance door, which will have an X-09 combination lock along with access control systems:

All SCIF perimeter doors must be plumbed in their frames and the frame firmly affixed to the surrounding wall. Door frames must be of sufficient strength to preclude distortion that could cause improper alignment of door alarm sensors, improper door closure or degradation of audio security. All SCIF primary entrance doors must be equipped with an automatic door closer.[1]

Doors will be solid core 1¾ inches thick; hinges should be within the secured side or nonremovable hinges if exposed (Figure 14.2).

Overly Doors and Krieger Doors both make excellent SCIF doors.

Windows

If you are at ground level or are less than 18 feet above the ground, there will be specific requirements. Windows will require protection against forced entry, vision, and sound attenuation.

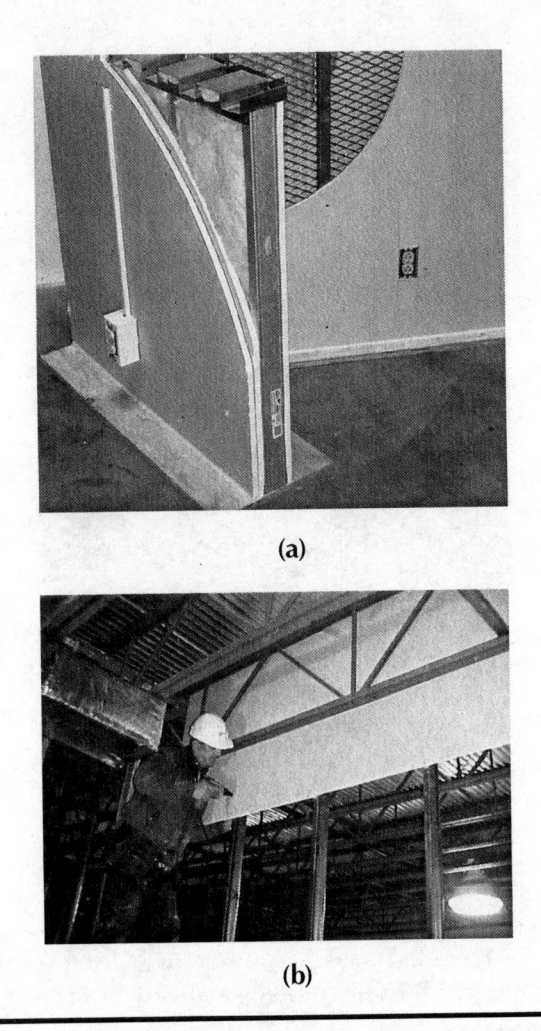

Figure 14.1 From left to right: (a) SCIF wall, (b) drywall installation, (c) SCIF corner and (d) sealing with caulk. (Photos courtesy of Project Developers, Inc, http://project-developers.com.) (continued)

Options are as follows:

- Frame in the window—simply put, you hang drywall right over the window and seal it up
- Apply blast-resistant film
- Install glass break sensors
- Drawn blinds with wands removed
- Install white noise window masker

(c)

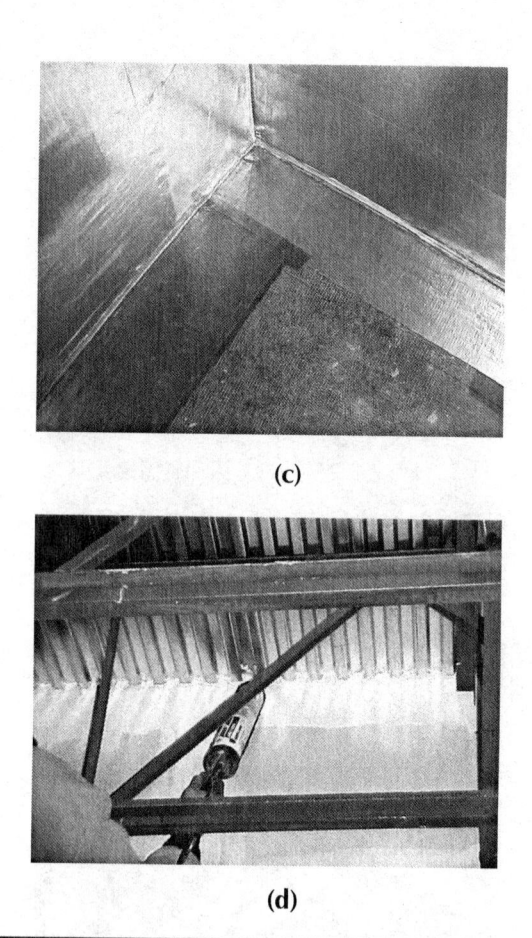

(d)

Figure 14.1 (continued) From left to right: (a) SCIF wall, (b) drywall installation, (c) SCIF corner and (d) sealing with caulk. (Photos courtesy of Project Developers, Inc, http://projectdevelopers.com.)

Electrical and HVAC

The electrical panel feeding the SCIF should be inside the SCIF. Power line penetration into the SCIF should be minimized. Sealing the conduit inside the protected area will be required. This can be accomplished with expandable foam that is typically used for insulation.

Basic HVAC (Figure 14.3a–d) requirements call for any duct penetration into the secured area that is over 96 square inches to require man bars so as not to allow a perpetrator to climb through the ducts. The exception is if the height of the duct is 6 inches or less and the remaining duct equals 96 square inches; then, man bars will not be required.

Most modern buildings use an open plenum return system instead of a duct return. In a true floor to true ceiling design, the HVAC engineer will

Figure 14.2 SCIF door with hardware and alarm panels. (Photo courtesy of Project Developers, Inc, http://projectdevelopers.com)

need to design plenum returns to allow stale air extraction. These can be as big as 2 feet by 3 feet. One method of protection would be to use a "Z" duct inside the protected area over the plenum. Another common method of protection is the installation of "man bars," which are ½-inch diameter steel and placed to allow the airflow but to keep anyone from using an opening bigger than 96 square inches to enter. The bars must be welded in place 5 inches on center.

An inspection port is also required at the wall penetrations from SCIF to non-SCIF. The inspection port is typically located on the SCIF side of the penetration. The port is typically located on the secure side so that no one from the nonsecure side can tamper with the port. It is acceptable, however, to locate the port on the nonsecure side as long as it is locked. Most inspection ports installed are hinged access doors.[2]

Sound Masking

Most sound masking is used to provide confidentiality in closed offices from a casual listener outside the office. The primary concern for why sound masking is needed beyond the physical construction is that most major strategic and tactical decisions are first made orally at meetings and, if an eavesdropper can obtain access, it gives them a distinct advantage over written or computer documents.

Figure 14.3 Above from left to right: (a) Z duct, (b) canvas break, (c) ceiling and drywall detail, and (d) man bars. (Photos courtesy of Project Developers, Inc, http://projectdevelopers.com) (continued)

White-noise or sound-masking devices need to be placed over doors, in front of the plenum, or pointed toward windows to keep an adversary from listening to classified conversations. Some SCIFs use music or noise that sounds like a constant flow of air to mask conversation.

Sound Classification

Sound transmission class (STC) is used in the discussion of architectural acoustics to describe the transmission attenuation afforded by various wall materials and other building components. The following STCs satisfy the normal security requirements of facilities used for SCI activities:

a. STC of 30 or better: loud speech can be understood fairly well; normal speech cannot be easily understood.

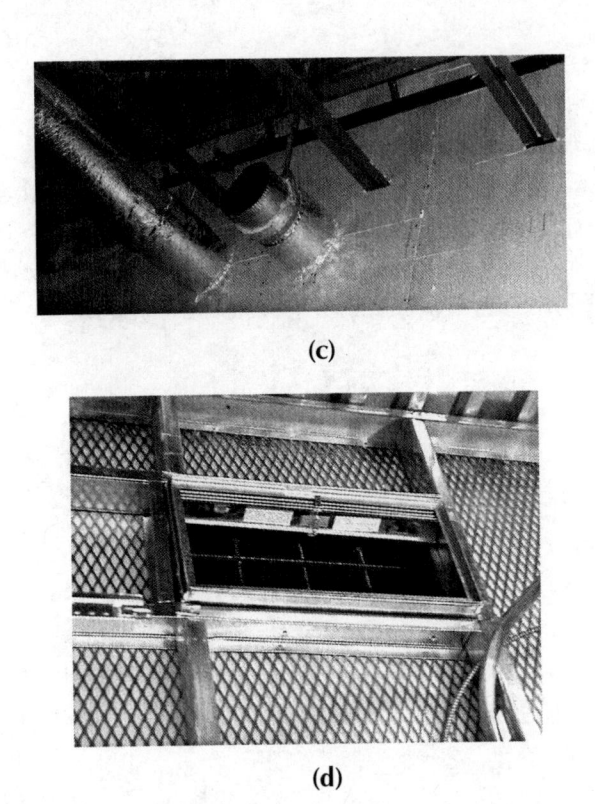

Figure 14.3 (continued) Above from left to right: (a) Z duct, (b) canvas break, (c) ceiling and drywall detail, and (d) man bars. (Photos courtesy of Project Developers, Inc, http://projectdevelopers.com)

b. STC of 40 or better: loud speech can be heard, but is hardly intelligible; normal speech can be heard only faintly, if at all.
c. STC of 45 or better: loud speech can be faintly heard but not understood; normal speech is inaudible.
d. STC of 50 or better: very loud sounds, such as singing, brass musical instruments, or a radio at full volume, can be heard only faintly or not at all.[3]

TEMPEST

It is a government requirement that all new SCIF construction and all SCIF requiring recertification will have TEMPEST measures in place. TEMPEST is used to guard against the susceptibility of some computer and telecommunications devices to emit electromagnetic radiation (EMR) in a manner that can be used to reconstruct intelligible data.

Shielding

Shielding devices and/or rooms with metallic materials are a reliable way to reduce the leakage of emissions. Examples of use of inherent shielding of standard construction materials include aluminum-foil-backed gypsum board, aluminum-foil-backed insulating sheathing, metallic-clad siding, copper foils (normally used for vapor barriers), wire meshes, and sheet metal roofing.

Filtering

Inserting filters into communication interface cables or power supply cables to suppress emissions is also effective. However, these filters need to be inserted into all of the external interfaces and tuned to the emission frequency. Moreover, this method is effective only if the emissions are radiating mainly from the interface lines rather than from the PC itself.

Access

Only personnel who possess appropriate SCI access authorization(s) and have been specifically authorized access may enter the SCIF unescorted. An access list of such personnel shall be maintained in a current status and posted immediately inside the entrance of the SCIF. Even before authorized personnel enter the SCIF, they must be made aware that portable electronic devices are *not* permitted inside.

Essential access to the SCIF by custodial, maintenance, or other uncleared persons shall be strictly controlled. Such persons will be closely escorted within the SCIF by individuals who possess the appropriate clearances. Prior to admittance of uncleared people, all occupants of the SCIF will be specifically advised of the presence of uncleared persons and all classified activities are to be suspended and classified material stored during the period when uncleared persons are present. This notification is typically done with a rotating light attached to the wall or ceiling. It can be in any color as long as personnel inside the SCIF know that, when it is activated, proper precautions are made while uncleared personnel are within the SCIF.

All individuals not assigned to, who do not work in, or who are not listed on the area access list posted within the SCIF must sign in on a visitor log, which contains a printed name, signature, date, time in, time out, and name of escort. Visitor logs for these approved areas must be retained for 5 years. Visitor logs may be locally produced.

All access control must be controlled from within the SCIF. This means a stand-alone access control system can be utilized with authorized personnel enrolled into the access control system. The door will maintain an access control system with at least two types of technology (badge, PIN, biometric, etc.). A standard card reader with a PIN will be sufficient for workday entry.

Security Alarm Requirements

The SCIF shall be fully secured (all classified material properly stored and perimeter door(s) locked and alarmed) during all periods whenever the area is not occupied by authorized personnel. The SCIF will be protected by an intrusion detection system (IDS) when not occupied. Intrusion detection is sent out to a central station with the requirement that a response force respond to the perimeter of the SCIF within 15 minutes.

The following are IDS requirements for the SCIF:

- All perimeter doors will be equipped with high-security balanced magnetic door switches.
- The interior spaces not continually occupied by authorized personnel will be protected by motion detection alarms.
- Areas above false ceilings will be alarmed in facilities that have been approved for open storage of SCI. Areas above false ceilings in other facilities will be alarmed if there is a possibility of surreptitious entry into the area above the false ceiling.
- All vents and ducts that penetrate a facility's perimeter and other openings with both dimensions over 6 inches will be alarmed.
- Windows less than 18 feet from ground level, or otherwise accessible from the roof, trees, or by other convenient methods, will be alarmed to prevent entry without detection.
- All alarm control units such as the arm/disarm keypads must be physically located within the secured perimeter of the area that the alarms are protecting and equipped with tamper switches.

- A minimum of 4 hours' standby power is required on all systems, except when a system is connected to a local uninterrupted emergency power source.
- Alarm equipment that is installed on perimeter walls, above the drop ceilings, vents/ducts, windows, emergency doors, etc., should remain in the "on" or "secure" mode at all times, even when the area is occupied.
- If all intrusion detection alarm equipment is installed on the same system (a single transmission line pair), it will be wired so that only the entrance door and interior alarm equipment (i.e., motion detection) are shunted out of the system when the access/secure (on/off) switch is placed in the "off" or "access" mode. Tamper switches must remain active at all times.[4]

SCIF entrance door-locking devices will typically be a CDX-09 lock. Access to the combination of the built-in combination security lock will be limited to the minimum number of individuals required to open the area. All combinations will be changed every 6 months or whenever someone who had access no longer requires access or whenever it is suspected the combination has been compromised.

The alarm system must be monitored by a UL-certified central station and all contractor-operated SCIFs shall maintain a current UL certificate of installation and service.[5]

Open Storage

An open storage area is used when the volume or bulk of classified material is such that the use of security containers is not practical. Open storage of classified material up to and including "secret" may be authorized within rooms that meet the vault or vault-type room requirements, including alarm protection and XO-series combination locks on the door. All classified material must be secured during nonworking hours in approved security containers or vaults.

According to Jeff Bennett,

> Ensure you have a classified contract that approves classified storage and performance at the prospective closed area location. You can find this information on the top right corner of the DD Form 254. There are two blocks there that indicate Facility Clearance

Required and Level of Safeguarding Required. Block 11 should be marked with the Cleared Contractor's requirements in performance of the classified contract (store, receive only, fabricate, etc.). Further instructions may be found in Blocks 13 and 14. If you have any questions, you should clear it up with the customer. Your responsibility as FSO is to ensure your company is capable of understanding the security requirements and performing as instructed. It is vital that your executives and customers are in complete synchronicity. Work with your Defense Security Services (DSS) to ensure they understand the requirements and there are no surprises. DSS has oversight and as such, they will verify that your classified contract, storage capability, and security program will protect classified information. As such, the cleared defense contractor, your organization, will also have to produce and demonstrate storage and performance procedures before approval.[6]

Closed Storage

SCI must be stored in GSA-approved containers having a resistance to surreptitious entry equal to or exceeding that of a class 6 container. There are no special construction requirements, so long as the floors, walls, and ceilings are constructed to substantial, permanent material that provides protection from surreptitious entry and will offer visual evidence of surreptitious or forced entry.

Dealing with Contractors

Having an understanding of SCIF construction will help prior to bringing on a contractor to build or renovate your SCIF. When looking for a quality contractor, you should find one that will provide preconstruction services to help you determine the scope and nature of your SCIF needs, as well as preliminary cost proposals for your SCIF construction project. The contractor should be able to lay the groundwork for your SCIF project by determining the type, size, configuration, and budget that meets your needs.

DO NOT start installing alarms, doors, locks, filters, etc., until you have received approval of your preconstruction fixed facility checklist (FFC; see Appendix A). The preconstruction FFC is your blueprint and the primary document in the decision-making process for building and getting your SCIF accredited. Your contractor will need to assist you in the entire accreditation process. He or she will need to be able to follow all government regulations for SCIF construction, including ICD 705, JAFAN requirements, conceptual plans, construction security plans, fixed facility checklist, and all other requirements.

References

1. DCID 6/9 3.3.3.2.
2. SCIF Info, Project Developers. http://www.projectdevelopers.com/scif.html
3. Sound Transmission Class (STC). http://www.acousticalsurfaces.com/acoustic_IOI/101_23.htm
4. DCID 6/9 3.3. Requirements common to all SCIFs.
5. UL Alarm Certificate of Compliance. http://qs.phly.com/selectsurvey/Pages/USRowing/docs/ALARMS_1100.pdf
6. Bennett, J. 2011. *Classified storage approval—Three steps to prepare defense contractors for closed areas.* Red Bike Publishing. http://industrialsecurity.wordpress.com/2011/08/17/classified-storage-approval-three-steps-to-prepare-defense-contractors-for-closed-areas/

Appendix A: Fixed Facility Checklist

Date _____

Fixed Facility Checklist

[ ] Preconstruction [ ] New facility [ ] Modified facility

Section A. General Information

1. SCIF data: organization/company name: _____
   SCIF identification number (if applicable): _____
   Organization subordinate to (if applicable): _____
   Contract number and expiration date: _____
   CSA: _____
   Project headquarter security office (if applicable): _____
2. SCIF location: _____
   Street address: _____
   Bldg. name/no.: _____ Floor: _____
   Room(s) no: _____
   City: _____ State/country: _____
   ZIP code: _____
3. Responsible security personnel:
   Primary: _____ Alternate: _____
   Commercial telephone: _____
   DSN telephone: _____
   Secure telephone: type: _____
   Home telephone: _____
   Fax no: (specify both classified and unclassified)
   Classified: _____ Unclassified: _____
   Other: _____
4. Accreditation data:
   a. Category of SCI requested: _____
      Indicate the storage required:
      ____ Open storage ____ Closed storage ____ Continuous operation
      ____ Secure working area ____ Temporary secure working area
   b. Existing accreditation information (If applicable):
      (1) Category of SCI: _____
      (2) Accreditation granted by: _____ on _____

c. Last TEMPEST accreditation (if applicable): accreditation granted by: _____ on _____
d. If automated information systems (AISs) are used, has an accreditation been granted?
____ YES ____ NO
Accreditation granted by: _____ on _____
e. SAP co-located within SCIF?____ YES ____ NO
If YES, classification: _____, and provide copy of co-utilization agreement for SAP operation in SCIF.
f. Duty hours: ____ hours to hours, ____ days per week.
g. Total square feet SCIF occupies: _____

5. Construction/modification: Is construction or modification complete?
____ YES ____ NO ____ N/A
If NO, expected date of completion: _____

6. Inspections:
   a. TSCM service completed by ____ on ____ (Attach copy of report.)
   b. Were deficiencies corrected? ____ YES ____ NO ____ N/A
   If NO, explain: _____
   c. Last physical security inspection by _____ on _____ (Attach copy of report.)
   Were deficiencies corrected? ____ YES ____ NO ____ N/A
   If NO, explain: _____
   d. Last security assistance visit by _____ on _____

7. Remarks: _____

Section B. Peripheral Security

8. Describe building exterior security:
   a. Fence: _____
   b. Fence alarm: _____
   c. Fence lighting: _____
   d. Television (CCTV): _____
   e. Guards: _____
   f. Other: _____

9. Building:
   a. Construction type: _____
   b. Describe access controls:_____
      (1) Continuous: ____ YES ____ NO
      (2) If NO, then during what hours? _____

10. Remarks: _____
   _____

Section C. SCIF Security

11. How is access to the SCIF controlled?
    a. By guard force: ____ YES ____ NO (Security clearance level: ____)
    b. By assigned personnel: ____ YES ____ NO
    c. By access control device: ____ YES ____ NO
       If YES, manufacturer ____ model no. ____
12. Does the SCIF have windows? ____ YES ____ NO
    a. How are they acoustically (if applicable)? _____
       _____
    b. How are they secured against opening? _____
       _____
    c. How are they protected against visual surveillance (if applicable)?
       _____
       _____
13. Do ventilation ducts penetrate the SCIF perimeter? ____ YES ____ NO
    a. Number and size (indicate on floor plan): _____
       _____
    b. If over 96 square inches, type of protection used:
       (1) IDS: ____ YES ____ NO (Describe in Section E)
       (2) Bars/grilles metal baffles: ____ YES ____ NO ____ OTHER
           (Explain: _____)
    c. Metal duct sound baffles: Are ducts equipped with:
       (1) Metal baffles: ____ YES ____ NO
       (2) Noise generator: ____ YES ____ NO
       (3) Nonconductive joints: ____ YES ____ NO
       (4) Inspection ports: ____ YES ____ NO
           If YES, are they within the SCIF? ____ YES ____ NO
           If they are located outside the SCIF, how are they secured?
           _____
    d. If TEMPEST accreditation authority is required, are pipes, conduits, etc., penetrating the SCIF equipped with nonconductive unions at the point they breach the SCIF perimeter?
       ____ YES ____ NO
       Are they provided acoustical protection (if applicable)?
       ____ YES ____ NO

14. Construction:
    a. Perimeter walls:
       (1) Material and thickness: _____
       (2) Do the walls extend from the true floor to the true ceiling?
           ____ YES ____ NO
    b. True ceiling (material and thickness): _____
    c. False ceiling? ____ YES ____ NO
       If YES:
       (1) Type of ceiling material: _____
       (2) Distance between false and true ceiling: _____
    d. True floor (material and thickness): _____
    e. False floor? ____ YES ____ NO
       If YES, distance between false and true floor: _____
15. Remarks: _____
    _____

Section D. Doors

16. Describe SCIF primary entrance door (indicate on floor plan):
    _____
    Is an automatic door closer installed? ____ YES ____ NO
    If NO, explain: _____
17. Describe numbers and types of doors used for SCIF emergency exits and other perimeter doors (indicate on floor plan): _____
    Is an automatic door closer installed? ____ YES ____ NO
    If NO, explain: _____
18. Describe how the door hinges exterior to the SCIF are secured against removal (if in an uncontrolled area): _____
    _____
19. Locking devices:
    a. Perimeter SCIF entrance door:
       (1) List manufacturer, model number, and group rating: _____
           _____
       (2) Does entrance door stand open into an uncontrolled area?
           ____ YES ____ NO
           If YES, describe tamper protection: _____
    b. Emergency exits and other perimeter doors:
       Describe (locks, metal strip/bar, deadbolts, panic hardware): _____
       _____

c. Where are the door lock combinations filed? _____
_____

20. Remarks: _____
_____

Section E. Intrusion Detection Systems

Give manufacturer and model numbers in response to the following questions:

21. Method of interior motion detection protection:
    a. Accessible perimeter? _____
       Storage areas? _____
    b. Motion detection sensors (Indicate on floor plan.): _____
       Tamper protection:____ YES ____ NO
    c. Other (e.g., CCTV, etc.): _____
22. Door and window protection (Indicate on floor plan.):
    a. Balanced Magnetic Switch (BMS) on door? _____
       Tamper protection: ____ YES ____ NO
    b. If SCIF has ground floor windows, how are they protected? ____
    c. Other (e.g., CCTV, etc.) _____
23. Method of ventilation and duct work protection: _____
    _____
24. Space above false ceiling (only outside the United States, if required):
    a. Motion detection sensors: _____
       Tamper protection: ____ YES ____ NO
    b. Other (e.g., CCTV): _____
25. Space below false floor (only outside the United States, if required):
    a. Motion detection sensors: _____
       Tamper protection: ____ YES ____ NO
    b. Other (e.g., CCTV): _____
26. IDS transmission line security protection:
    a. Electronic line supervision (manufacturer and model): ____
       If electronic line supervision: class of service: ____ I ____ II
    b. Other: _____
27. Is emergency power available for the IDS? ____ YES ____ NO
    Type: ____ Battery ____ Emergency generator ____ Other
28. Where is the IDS control unit for the SCIF located? (Indicate on floor plan.)
    _____

29. Where is the IDS alarm enunciator panel located? (Indicate on floor plan.) Address? _____

30. IDS response personnel: Describe: _____
    Response force security cleared: ____ YES ____ NO
    a. Level: _____
    b. Emergency procedures documented? ____ YES ____ NO
    c. Reserve force available? ____ YES ____ NO
    d. Response time required for alarm condition: _____ minutes
    e. Are response procedures tested and records maintained?
       ____ YES ____ NO
       If NO, explain: _____
31. Is the IDS tested and records maintained? ____ YES ____ NO
    If NO, explain: ____
32. Remarks: _____

Section F. Telephone System

33. Method of on-hook security provided:
    a. TSG-2 computerized telephone system (CTS)? ____ YES ____ NO
       (1) Manufacturer/model: _____
       (2) Location of the CTS: _____
       (3) Do the CTS installers and programmer have security clearances?
           _____
           If YES, at what access level (minimum established by CSA):____
           If NO, are escorts provided? _____
           _____
       (4) Is the CTS installed per TSG-2 configuration requirements?
           ____ YES ____ NO
           A. If NO, provide make and model number of telephone equipment, explain your configuration, and attach a line drawing.
              _____
           B. Is access to the facility housing the switch controlled?
              ____ YES ____ NO
           C. Are all lines between the SCIF and the switch in controlled spaces?
              ____ YES ____ NO

(5) Does the CTS use remote maintenance and diagnostic procedures or other remote access features? ____ YES ____ NO
If yes, explain those procedures: _____
_____

b. TSG-6 approved telephones?
  (1) Manufacturer/model: _____
  (2) TSG number: _____
  (3) Ringer protection (if required): _____
c. TSG-6 approved disconnect devices?
  (1) Manufacturer/model: _____
  (2) TSG number: _____

34. Methods of off-hook security provided:
  a. Is there a hold or mute feature? ____ YES ____ NO
    (1) If YES, which feature _____? Is it provided by the CTS or _____ telephone?
    (2) If NO, are approved push-to-operate handsets provided?
       ____ YES ____ NO
       If YES, describe: _____

35. Automatic telephone call answering:
  a. Is there an automatic call answering service for the telephones in the SCIF?
  b. ____ YES ____ NO
    If YES, provide make and model number of the equipment, explain the configuration, and provide a line drawing. _____

Section G. Acoustical Protection

36. Do all areas of the SCIF meet acoustical requirements?
  ____ YES ____ NO
  If NO, describe additional measures taken to provide minimum acoustical protection (e.g., door, windows, etc.) _____

37. Is the SCIF equipped with a public address, emergency/fire announcement, or music system? ____ YES ____ NO
  If YES, describe and explain how it is protected: _____
_____

38. If any intercommunication system that is not part of the telephone system is used, describe and explain how it is protected: _____
_____

39. Remarks: _____
   _____

Section H. Administrative Security

40. Destruction methods:
    a. Describe method used for destruction of classified/sensitive material:
       Manufacturer: _____ Model: _____
       Manufacturer: _____ Model: _____
    b. Describe location of destruction site(s) in relation to the secure facility:
       _____
    c. Have provisions been made for the emergency destruction of classified/sensitive program material? (If required): ____ YES ____ NO
       If YES, has the emergency destruction equipment and plan been coordinated with the CSA? ____ YES ____ NO

41. If reproduction of classified/sensitive material takes place outside the SCIF, describe equipment and security procedures used to reproduce documents:
    _____

42. Remarks: _____
    _____

Chapter 15

Financial Institutions and Banks

Paul R. Baker

Introduction

The concepts of deter, detect, deny, and respond are the same physical security approaches that are applied to a bank. When dealing with any institution or operation, the same functional security model can be relevant and effective. As Willie Sutton purportedly stated when asked why he robbed banks, "Because that's where the money is." Knowing that a bank is a target for robberies, it is essential that protective measures be employed to protect employees, customers, and assets.

Vaults

A bank vault is a secure space where money, valuables, records, and documents can be stored (Figure 15.1). It is intended to protect contents from theft, unauthorized use, fire, natural disasters, and other threats. Most banks typically maintain their safety deposit boxes within a vault and use them as places for tellers to keep their cash drawers. During construction of a new bank, a vault is the first thing to be placed within the floor plan of the space

Figure 15.1 Bank vault.

and the building is constructed around the vault. Bank vaults are typically made with reinforced steel.

The vault door should be protected with contact switches that are set off by opening the door. Vault alarms should be scheduled to be opened and closed to coincide with banking hours. Typically, a vault will have an array of alarm systems to include sound, motion, heat/thermal changes, and seismic vibration.

Safes

Safes come in all sizes and shapes, but the difference between a safe and a vault is that you can walk inside a vault and a safe is a secured container. Safes can be used for holding teller drawers and additional cash. They are also used in conjunction with night deposits, ATMs, and cash dispensers. According to UL, the minimum approved safe for the banking industry will have a TL-15 rating.

Tool-resistant safe class TL-15. This type of combination-lock safe is designed to meet the following requirements: It must be resistant to entry for a networking time of 15 minutes using any combination of the following tools: mechanical or portable electric hand drills not exceeding ½-inch size, grinding points, carbide drills (excluding the magnetic drill press and other pressure-applying mechanisms, abrasive wheels, and rotating saws), and common hand tools such as chisels, drifts, wrenches, screwdrivers, pliers, and hammers and sledges not to exceed the 8-pound size; pry bars and ripping tools not to exceed 5 feet in length; and picking tools that are not specially designed for use against a special make of safe.[1] A certified UL safe will have an official tag attached to the safe with identification of what level of protection it will provide and a serial number from UL.

ATMs

An automated teller machine (ATM) is a computerized telecommunications device that provides the customer with access to financial transactions in a public space without the need for a teller. All ATMs require a plastic credit card with a magnetic stripe or a smart card with a chip. Authentication is provided by the customer entering a personal identification number (PIN). This allows the customer to withdraw cash, check balances, or make deposits.

ATMs are either within the confines of a bank branch or are stand-alone machines at off-site locations. The ATM is bolted into the concrete floor and typically weighs 250 pounds, depending on the model. The device is protected with a built-in safe that has door contact and heat/vibration sensors. The top portion of the ATM is called the hood, which will also have a contact switch for protection from unauthorized entry. The alarm system will be connected via the network or phone line or both.

ATMs maintained within the branch proper have better protection just from the fact that they are within the layers of security provided to a bank (Figure 15.2). Off-site ATMs should be located within a business location such as a grocery store, drug store, or locations where there is adequate protection.

Figure 15.2   Bank branch ATM.

Using video recording of ATM transactions is typical for branch ATMs. The DVR or NVR for the branch can also hold the ATM video. It is important that a transaction-capture device be installed with the ATM video. This will allow fraud investigators quickly to identify transactions with video recordings. Off-site ATMs have more issues with the installation of cameras especially if they require a DVR for recording. The lack of room for a small two-camera DVR may work but it would be better for the installation of an IP camera to an NVR system.

Night Depositories

A night deposit or night drop (Figure 15.3) is a device where merchants deposit their daily receipts after hours. The night deposit will be installed within the secured portion of the back with its shoot opening exposed to

Figure 15.3 Night deposit.

the street. This can be accessed only by authorized merchants who have a relationship with the bank and who have been issued a key to open the shoot. The protection measures for a night deposit should consist of a wire lacing around the body of the depository head and the connecting chute. This will raise an alarm if someone is trying to fish out a bag with a wire. The chest door should be fitted with a contact switch and the entire component should have heat/vibration alarm sensors.

Teller Cash Recyclers

Cash recyclers are relatively new to the banking industry. They are used for better cash control and to reduce fraud, eliminate cash exposure, and provide for better physical security. "This means that banknotes that are deposited at the counter in a bank branch and stored in intelligent cassettes in an automated safe can be made available again for dispensing on automated teller machines."[2]

Cash recycling allows for faster, more accurate cash transactions and reduced customer wait times. Cash recycling can dramatically reduce teller balancing time. Tellers can balance at the end of a shift in less time than it would take without a cash recycler while eliminating balancing errors. A recycler has all the advantages of a currency counter/discriminator that can count the value of a deposit, detect unfit and counterfeit notes, and sort deposits by denomination into a safe.

Dye Packs

Dye packs are unique to the banking business. They are security devices that contain either dye or tear gas or both. The dye is generally a bright red color that cannot be removed from currency or clothing. 3SI Security is the primary manufacturer of dye packs, which all generally operate on the same principle; that is, the dye packs resemble currency, and the dye or tear gas is inserted within a hollowed-out stack of money and is maintained in a teller's working cash drawer. When the dye pack is moved off its protective magnet, located in the drawer, it activates the battery and when it is taken out of the bank, a radio frequency transmitter placed at the exit activates a receiver in the dye pack that triggers the release of the dye or tear gas. This release can be adjusted to various time frames after the robber leaves the banking office.

Some security officers, however, have expressed concern about the accidental discharge of a dye pack within the banking office and have decided not to put them into use. In this instance, a product known as a "Scoop and Go" is advisable. It is a hollowed-out pack and will have real denominations attached to the top and bottom to give the impression of a large pack of money. When a robber demands money from the teller, this bait money pack can be "scooped" up out of the teller drawer and given to the robber. At the most, the robber would be given around $100 while thinking that he or she has received several thousand dollars. Also, the Scoop and Go is a onetime cost (approximately $45 per pack) and can be kept in circulation for a very long time. This is an inexpensive way to maintain bait money. Typically, a dye pack can cost up to $400 and will need to be replaced every 2 years.

The latest and greatest advance in bank robbery devices is the use of a tracking system that is embedded within the pack. This is similar to GPS or a LoJack type of system. The local police department must be a part of the network in order to track the device, but overall it is a very effective way to track and arrest the bank robbers within minutes of a bank robbery. However, this device is the most expensive of the three styles of bait money and also includes a yearly fee for the transmitter.

One of the selling points for using bait money and dye packs is that 3SI provides a reward for up to the installation cost when a dye pack is given out by the teller.

Barriers

Bullet-resistant glass, or BR glass, is used for the protection of the teller (Figure 15.4). It is not a deterrent to bank robberies other than as a psychological effect and a sense of protection for the teller staff. The standard installation of BR glass is at level 1, which can stop up to a .9mm round. It is ineffective after several shots and the rounds can penetrate. It is designed only to protect the tellers; there is no protection for the lobby staff or customers. Typically, in a bank with five teller windows, the cost for level 1 BR glass can average around $50,000. This needs to be weighed against the risk analysis for the particular branch.

Figure 15.4 Bullet-resistant glass at a bank.

Ambush Features

One of the main concerns for employees entering a branch in the morning is the possibility of a "morning glory attack" or an ambush. An assailant will wait for a branch employee to unlock the branch in the morning, ambush the employee, and have him or her open the branch and the safe. There is also the possibility of an assailant concealing himself or herself inside the branch and waiting for an employee to open up the branch.

For this scenario, alarm systems have been preprogrammed for an ambush command. When the employee puts in a specific code (known only to the branch personnel), it will disarm all the devices as normal but will also send a silent alarm to the central alarm station. Also, the majority of banks have a two-person rule requiring the first person to enter and inspect the branch before placing or giving the all-clear signal for the second employee to enter.

Video Systems

CCTV systems are prevalent in the banking industry. They are positioned to capture images of robbery suspects, customer transactions, customer fraud, and employee theft. CCTV has also been incorporated into the alarm system to give visual alerts to central station operators for holdup verification or after-hours entry. Some banks use a video monitor positioned at the entrance that will show the image back to the customer as he or she walks in. This will let any would-be robber know that his or her image has already been captured by the bank's CCTV system.

Holdup Alarms

Typically, if you are in the business of providing security for a financial institution, you will be robbed—It's just a fact of life. The Bank Security Act requires *an alarm system or other appropriate device for promptly notifying the nearest responsible law enforcement officers of an attempted or perpetrated robbery or burglary.*[3] Use of holdup alarm devices by banking office employees provides a means to silently notify law enforcement of the robbery attack. There are typically three types of bank robberies:

1. Note passers—a robber who passes a note to the teller demanding money and who may or may not state that he or she has a weapon.
2. Armed robber—a robber who brandishes a weapon and demands money from the teller. Federal Bureau of Investigation (FBI) statistics generally show that in about one-half of the robberies committed against banks, the robbers use guns during commission of their attacks.
3. Takeover robber—an armed robber who gets behind the teller line and forces bank personnel to open the safe. This is typically the most violent of the three.

In any robbery scenario, it is a requirement to notify local law enforcement. There are three common devices for sending a silent alarm: holdup buttons, foot rails, and money clips. These devices will sound an alarm at a central alarm station, which will contact the branch to verify that a robbery has occurred. It is imperative that there be verification prior to dispatching the police.

Holdup Buttons

A holdup button is an inconspicuous, hand-activated wire device that operates silently to trigger a silent robbery alarm system. The buttons are usually depressed in the unit to avoid accidental triggering of the device. Holdup buttons and other holdup devices should be mounted under all tellers' counter stations. It is also recommended that holdup buttons be placed in the knee space at one or more desks in the customer service area.

Holdup Foot Rails

A holdup foot rail is a foot-operated alarm device installed on the floor of the teller's booth that can be used to activate the robbery alarm system.

Money Clips

Money clips are used in cash drawers and provide tellers with a safe and inconspicuous method of signaling that a holdup is taking place. They are activated when a packet of currency, normally bait or marked money, is removed from the clip.[3]

Duress alarms are semiautomatic in operation and can be placed on combination locks used to open vaults, safes, and cash storage lockers inside vaults. When an employee dials a predetermined incorrect combination, a silent duress alarm is activated, thus enabling branch personnel under duress to send an alarm discreetly to the central station or law enforcement while opening the storage container.

Bank Guards

In all the movies, you see a bank guard standing in the lobby of the bank. And in those movies, he is usually the first person to be in the line of fire and shot at. So what is the purpose of having an armed guard inside a bank? Financial institutions today use fewer guards than were used 25 years ago. Several factors account for this reduction in guard use, including:[3]

- Restraints on the use of deadly physical force by private security personnel
- Increasing costs attendant to the use of guards
- Development of sophisticated alarm and surveillance systems and electronic access-control devices that can replace guards
- Concerns by bankers that the use of guards in banking offices could lead to the accidental injury of customers and employees

It is not that bank guards should never be used, but that it should be done with a developmental plan rather than a "this is the way we have always done it" attitude (Figure 15.5). Initiating a risk assessment would be the proper process to determine whether a bank location was a higher or lower security risk. Each factor would be assigned a rating of 1 (low

Figure 15.5 Guards should be used with a specific developmental plan and purpose in place.

risk) to 3 (high risk). This risk assessment checklist would summarize the research that had been done at each location and include factors such as the following:

- **Historical losses**—the number of robberies or other serious security problems the office has had in the past
- **Crime rate data**—to determine the current level of security risk at the office
- **Neighborhood risk**—determined from the census data and site visits
- **Employee opinions**—an overall assessment based on the results of questionnaires on whether the location needs a guard

Risk factor ratings would then be added together to develop a guard-needs score for each location; these scores range from 6 to 20, indicating low risk to high risk, respectively. It would be up to management to determine at what point on the range a guard would be required.

References

1. Allied Safe and Vault Co. UL ratings. http://www.alliedsafe.com/ul_ratings.php#TL15
2. Lowe, F. 2010. ATM Marketplace. Wincor Nixdorf introduces a cash-recycling teller to the U.S. market. http://www.atmmarketplace.com/article/127485/Wincor-Nixdorf-introduces-a-cash-recycling-teller-to-the-U-S-market
3. *Bank Security Desk Reference.* 2009. Version 9.1. Washington, DC: A. S. Pratt & Sons.

Chapter 16

Data Center Protection

Paul R. Baker

Data Centers

When discussing the need to secure the data center, security professionals immediately think of sabotage, espionage, or data theft. While the need is obvious for protection against intruders and the intentional harm caused by intentional infiltration, the hazards from ordinary activity of personnel working in the data center present a greater day-to-day risk for most facilities.

Personnel within the organization need to be segregated from access to areas where they have no "need to know" for that area. The security director will have physical access to most of the facility but has no reason to access financial or HR data. The head of computer operations might have access to computer rooms and operating systems, but not the mechanical rooms that house power and HVAC facilities. It comes down to: Do not allow wandering within your organization. If you were working in the data center and you saw the line cook from the cafeteria walking through, how good would you feel about security and the protection of information?

As data centers grow, the need for physical security at the facility is every bit as great as the need for cyber security of networks. The data center is the brains of the operation and as such only specific people should be granted access.

The data center should have signs at the door(s) marking the room as restricted access and prohibiting food, drink, and smoking in the computer room. There should be a mandatory authentication method at the entrance to the room such as a badge reader.

The NOC (Network Operations Center) is the central security control point for the data center. This is the internal gatekeeper for the data center. It must have fire-, power-, weather-, temperature-, and humidity-monitoring systems in place. The NOC must have redundant methods of communication with the outside to include a telephone, cell phone, or two-way radio system. The NOC must be manned 24 hours a day.

Access to the data center should be restricted to those who need to maintain the servers or infrastructure of the room. Service engineers must go to the NOC to obtain access to the data center, server room, or computer room.

Cleaning crews should work in groups of at least two. The cleaning crew should be restricted to offices and the NOC. If cleaning staff must access a data center for any reason, they should be escorted by NOC personnel.

The standard scenario for increased security at a data center would consist of the basic security in depth: progressing from the outermost (least sensitive) areas to the innermost (most sensitive) areas. Security will start with entry into the building, which will require passing a receptionist guard and then using a proximity card to gain building entry. For access into the computer room or data center, it will now require the same proximity card along with a PIN, plus a biometric device (Figure 16.1).

Combining access control methods at an entry control point will increase the reliability of access for authorized personnel only. Using different methods for each access level significantly increases security at inner levels, since each is secured by its own methods plus those of outer levels that must be entered first. This would also include internal door controls. For a data center, the use of an internal mantrap or portal would provide increased entry and exit control. A portal allows only one person in at a time and will only open the inner door once the outer door is closed. The portal can have additional biometrics within the device that must be activated before the secured side door opens.

According to Kevin Beaver, principal security consultant of Principle Logic LLC, there are 10 common mistakes companies make when it comes to the physical layout of their data center:[1]

1. **Weak or missing security policies:** Don't take the time to develop security policies only to put them on a shelf and forget about them. It's important to make sure security policies are effectively communicated to employees. A good security

Figure 16.1 Biometric reader for access control.

policy includes a simple introduction that conveys the purpose of the policy, the policy statement itself and information about how compliance will be measured. It should also include information about what sanctions will be taken against those that fail to comply.

2. **Poor physical access controls:** To be sure that everyone entering the data center has a reason to do so, implement strong visitor sign-in procedures and then enforce those rules. If keycards are required to enter the data center, check regularly to make sure they work. Companies that have no receptionist or a distracted receptionist should consider hiring guards around the clock.

3. **Specific security concerns:** Constantly check the data center for vulnerabilities. Look to see how many access points there are and if people tend to prop doors open. Don't leave media such as CD-ROMs and other documentation lying around. Try to make sure that wires are not exposed. For companies that outsource their data center,

make sure the third party secures documentation about your infrastructure.

4. **Location and layout:** There is much debate over which floor of an office building is best for housing a data center. First-floor data centers are vulnerable to car crashes, while second-floor data centers may be vulnerable to fires that start below. Either way, try to be aware of where your data center resides in the building and develop disaster recovery plans accordingly.

5. **Unsecured computers:** Always lock screens when employees get up and walk away from their computer. Locking screensavers are recommended.

6. **Utility weakness:** Confirm that the proper fire protection policies are in place. Also, make sure there are working backup generators or battery power in the event of an electrical outage.

7. **Rogue employees:** Everyone inside the data center should have a reason to be there. Don't assume someone is trustworthy just because [he or she has] gained access to the data center. To solve the problem of rogue employees, vendors, and others passing through the data center, refer to internal policies or create them if necessary. Next, have some awareness training for employees. Finally, make it a human resources (HR) issue.

8. **Separation of physical and logical security:** Physical and logical security should be converged into one because they are both equally important. After all, there is a lot of overlap between the two. Both require risk assessment and countermeasures to mitigate risks. The goal of both is to keep the bad guys out and the good guys honest.

9. **Outsourcing all data center security responsibilities:** Companies should never outsource 100% of their data centers' security responsibilities to a third-party company. Rather, put someone in charge of making sure the third party is properly handling your physical security, compliance, and other needs.

10. **No third-party security assessments and/or audits:** The security of data centers is a continually evolving process. Every time a new technology is introduced, a new vulnerability appears that needs to be addressed. That is why it's

important to occasionally bring in a third-party auditor or consultant. Companies that outsource data center operations should consider sending auditors to the third-party company in question.

Fire Protection

To protect your data center from fire, you need to have smoke detectors installed and linked to a panel with enunciators that will warn people that there is smoke in the room. Also, it should be linked to a fire suppression system that can help put out the fire with no damage to equipment from the chemical itself. The room itself should be built with fire-retardant material to decrease the chance of fire. The installation of limited combustible cabling (LCC) should be used. This product is jacketed and insulated with a fluoropolymer resin and can hang on for as long as 20 minutes before it begins to catch fire—up to 10 times longer than a typical communication metallic plenum-rated cable.

Fire Detection/Alerting

Fire detection and alarm systems are designed to provide warning of a fire outbreak and allow for appropriate fire fighting action to be taken before the situation gets out of control, resulting in serious damage to property and possible loss of lives.

The fire alarm panel (Figure 16.2) is the hub of the fire alarm system in a building. It is usually located on the ground floor near an entrance close to the nearest road. The fire alarm panel senses the presence of a fire by way of smoke and heat detectors. Each detector is linked back to the panel, advising of the relevant zone of the building where the fire has been detected. The panel will automatically notify the fire department of an alarm when one of its sensors locates a fire.

A smoke detector is one of the most important devices to have to warn of a pending fire, coupled with a good signaling device. A detector in proper working condition will sound an alarm and give all occupants a chance to make it out alive. There are two main categories of smoke detectors: optical detection (photoelectric) and physical process (ionization).

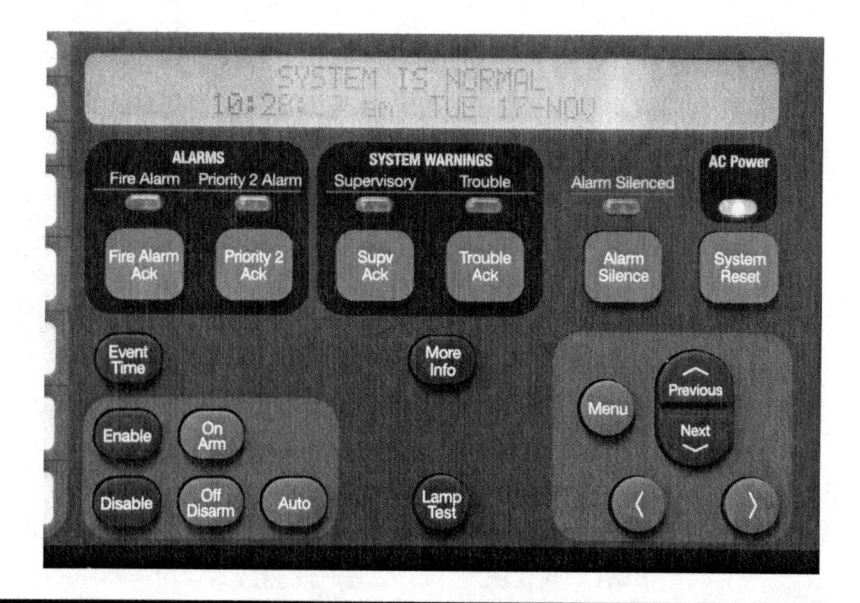

Figure 16.2 The fire alarm panel is located on the ground floor near an entrance close to the nearest road.

Photoelectric detectors are classified as either beam or refraction. Beam detectors operate on the principle of light and a receiver. Once enough smoke enters the room and breaks the beam of light, the alarm is sounded. The refraction type has a blocker between the light and the receiver. Once enough smoke enters the room, the light is deflected around the beam to the signal. The ionization type detector monitors the air around the sensors constantly. Once there is enough smoke in the room the alarm will sound.

The use of smoke detectors is very crucial; most deaths that occur in a fire are not from the fire or the flames. Most deaths are from smoke inhalation, heat, the toxic gases produced by fires, and explosion or panic. According to the National Fire Prevention Association,[2] in 2006 alone there were 3,245 deaths caused by fire, over 16,000 injuries due to fire, and over $11 billion in property damage. Nationwide, there is a civilian death due to fire every 162 minutes. With these statistics, we cannot afford to ignore the need for a proper fire execution plan.

As the old saying states, "If there is smoke, there must be fire." There are three main types of fire detectors: flame detectors, smoke detectors, and heat detectors. The two main types of flame detectors are classified as infrared (IR) and ultraviolet (UV) detectors. IR detectors primarily detect a large mass of hot gases that emit a specific spectral pattern in the location of the detector; these patterns are sensed with a thermographic camera and an alarm is sounded. Additional hot surfaces in the room may trigger a false response with this alarm. UV flame detectors detect flames at speeds of 3 to 4 milliseconds due to the high-energy radiation emitted by fires and explosions at the instant of their ignition. Some of the false alarms of this system include random UV sources such as lightning, radiation, and solar radiation, which may be present in the room.

There are heat detectors that include fixed-temperature or rate-of-rise detectors. This is a really simple concept for all users. The user will set a predetermined temperature level for the alarm to sound. If the room temperature rises to that setting, the alarm will sound. Rate-of-rise temperature will detect a sudden change of temperature around the sensor. Usually, this setting is at around 10° to 15° per minute. Nothing more is really required of the consumer except routine checks for battery life and operation status.

Heat detectors should not be used to replace smoke detectors; each component in fire safety serves its purpose and should be taken seriously. The combination of devices and the knowledge of procedures is the only way to achieve success during a possible fire.

Fire Suppression

Fire requires three elements to burn: heat, oxygen, and a fuel source. Fire extinguishers and fire suppression systems fight fires by removing one of the three elements. Fire extinguishers are divided into five categories, based on different types of fires:[3]

- **Class A** extinguishers are for ordinary combustible materials such as paper, wood, cardboard, and most plastics. The numerical rating on this type of extinguisher indicates the amount of water it holds and the amount of fire it can extinguish.
- **Class B** fires involve flammable or combustible liquids such as gasoline, kerosene, grease, and oil. The numerical rating for a class B extinguisher indicates the approximate number of square feet of fire it can extinguish.
- **Class C** fires involve electrical equipment, such as appliances, wiring, circuit breakers, and outlets. Never use water to extinguish class C fires because the risk of electrical shock is far too great! Class C extinguishers do not have a numerical rating. The C classification means the extinguishing agent is nonconductive.

- **Class D** fire extinguishers are commonly found in a chemical laboratory. They are for fires that involve combustible metals, such as magnesium, titanium, potassium, and sodium. These types of extinguishers also have no numerical rating and are not given a multipurpose rating; they are designed for class D fires only.
- **Class K** fire extinguishers are wet chemical discharge and are found in a restaurant kitchen environment. The class K extinguisher is hand portable and is the ideal choice for use on all cooking appliances including solid fuel char broilers.

All buildings should be equipped with an effective fire suppression system, providing the building with around-the-clock protection. All facilities should have portable fire extinguishing equipment conveniently located on each floor and especially within sensitive areas such as the data center. There is an acronym for using a fire extinguisher: PASS, which stands for

Pull the pin on the fire extinguisher.
Aim at the base of the fire.
Squeeze the handle on the extinguisher.
Sweep from side to side of the fire.

Traditionally, fire suppression systems employed arrays of water sprinklers that would douse a fire and surrounding areas. Sprinkler systems are classified into four different groups: wet, dry, preaction, and deluge:

- **Wet** systems have a constant supply of water in them at all times; once activated, these sprinklers will not shut off until the water source is shut off.
- **Dry** systems do not have water in them. The valve will not release until the electric valve is stimulated by excess heat.
- **Preaction** systems incorporate a detection system that can eliminate concerns of water damage due to false activations. Water is held back until detectors in the area are activated.
- **Deluge** systems operate in the same function as the preaction system except all sprinkler heads are in the open position.

Water may be a sound solution for large physical areas such as warehouses, but it is entirely inappropriate for computer equipment. A water spray can irreparably damage hardware more quickly than encroaching smoke or heat. Gas suppression systems operate to starve the fire of oxygen. In the past, halon was the choice for gas suppression systems; however, halon leaves residue, depletes the ozone layer, and can injure nearby personnel.

The following gas suppression systems are recommended for fire suppression in a data center/server room or anywhere electronic equipment is employed:

- Aero-K uses an aerosol of microscopic potassium compounds in a carrier gas released from small canisters mounted on walls near the ceiling. The Aero-K generators are not pressurized until fire is detected. The Aero-K system uses multiple fire detectors and will not release until a fire is "confirmed" by two or more detectors (limiting accidental discharge). The gas is noncorrosive, so it does not damage metals or other materials. It does not harm electronic devices or media such as tape or discs. More important, Aero-K is nontoxic and does not injure personnel.
- FM-200 is a colorless, liquefied compressed gas. It is stored as a liquid and dispensed into the hazard as a colorless, eclectically nonconductive vapor that is clear and does not obscure vision. It leaves no residue and has acceptable toxicity for use in occupied spaces at design concentration. FM-200 does not displace oxygen and, therefore, is safe for use in occupied spaces without fear of oxygen deprivation. The plus side of FM-200 is that it is safe. People can be in the room when it goes off and it is totally benign to electronic equipment. The downside is the cost.

References

1. Brunelli, M. 2004. Data center security: 10 things not to do. SearchCIO.com. http://searchdatacenter.techtarget.com/news/article/0,289142,sid80_gci1071551_tax305172,00.html
2. http://www.nfpa.org
3. http://www.fire-extinguisher101.com/

Chapter 17

Total System Cost

Daniel J. Benny

Determining Total System Cost

When determining the total security system cost, several categories must be explored:

- System design cost
- System installation cost
- System operational cost
- Maintenance cost
- IT cost
- Replacement cost
- Return on investment

It is important in the development of a security system that the total cost of the system be attained in order to develop a realistic budget that can be justified to top management and to ensure the system that is installed meets the security requirements of the organization based on the threat.[1]

System Design Cost

Initially there is the cost to develop the specifications for the project. During this phase, depending on the complexity and sophistication of the total security system, the assistance of a security consultant or engineer may be required.

During this phase the examination of the requirements would include the type of security system that would be the most effective based on the threat and location being protected and the various components of the system. These system components would include the intrusion detection system central station server. They would also include the computer that would be utilized to operate the security system. The monitors that will be required to work the system must also be included in the system design. Depending on the size and number of monitors, the construction of a rack system to hold the monitors may be required.[2]

The various types and the number and placement of the security sensors will need to be determined and documented. As an example, it would include electronic door contacts or passive infrared sensors. The number and placement of fire sensors to include smoke detectors, heat detectors, and water flow sensors also needs to be determined.

Access controls such as card readers and cyber locks and traditional locks and their placement in the facility must be identified. If electronic access control devices are utilized, conduit and wiring to power the units will need to be calculated into the cost of the project.

The number and operating requirements of the security cameras will need to be identified. This will include the type of lens, camera body, and operating aspects such as zoom capabilities and transmission method. Conduit and wiring to power the units will need to be calculated into the cost of the project, which will be integrated into the total security system that is being planned.[2]

The design cost will also include development of the drawings and blueprints of the total system that is to be constructed and installed. There are, of course, the consultant fees for the individual or firm that is hired to design the security system. Cost for the engineer or engineering firm that will create the drawing and blueprints of the project must also be considered in the design process.

There are many aspects of the system design cost that must be taken into account. This will be important when submitting a budget for such a project. The life cycle of the security system should also be a consideration for long-term budget projection.

System Installation Cost

One of the most expensive aspects of the entire security system project will be the system installation cost. This includes the cost of the products or components of the security system:

- Server
- Computer
- Monitors
- Control panel
- Wiring
- Metal conduit
- Security camera
- Camera brackets and housing

There is also the expense of the various sensors, such as door and window contacts, motion sensors, and fire protection sensors integrated into the system. If access control is part of the system, then there is the cost of the readers and cards to be used with the product.

Once the products have been identified and purchased, there will be the shipping cost to transport the system components to the installation site. This could include fees for rail and truck transport of large parts and the cost of local carriers for smaller products associated with the security system.

Labor costs for the individuals installing the system can be sizeable based on the local union or non-union wages in the local area. This would include electricians. If other construction is needed to support the security system, it may also include masons, carpenters, and painters.

In most cases, permits from the local government or municipality will be required for new construction and electrical installations. The cost of the permits will vary based on these local governments and their specific requirements for where the facility and the security system project are located. Based on the nature of the product, there may also be state or Environmental Protection Agency permit fees to pay.

System Operational Cost

Once the system is installed, there will be initial and ongoing system operation costs.

In order to ensure the proper function of the system, current policies will need to be rewritten as well as new policies written with regard to the operation of the security system. These operational changes may impact how other departments in the organization operate, causing additional cost to make such changes to the company operating policy and infrastructures.

Since all new security systems are computer based, there will be significant initial and ongoing support from the organization's IT department. This includes integrating the security system into the company IT system, the development of IT security procedures, and software to protect the system.

The increase in cost for electrical power is also part of the system operating cost. In the event of a power loss, the security system must function, so an emergency backup generator must be included in the ongoing cost.

The most expensive ongoing cost will include initial and continued training of the security staff and the wages for additional security staff that will monitor the security system. In some cases, the addition of a comprehensive security system may free some security officers on patrol to monitor the system, but this is not the norm. In most situations, additional security will need to be hired.

IT-Related Cost

When developing a new security system that is computer based, there will be IT-related costs. It is vital to know what IT systems are available on the corporate IP network. In the total coast of the security system you will need to account for cost factors associated with industry best practices for the management of IP-based technologies, such as:

- Antivirus technology
- System patches
- Database management
- Backup and archiving
- Network bandwidth and quality of service

Each of these costs will have an associated cost for labor and IT personnel, who may be dedicated partly or fully to monitoring and maintaining the new systems.

Maintenance Cost

Keeping the system operating will require an investment in ongoing maintenance. This would include routine costs to keep the system hardware running and upgrades to the software. It will also require updates to the physical components of the system such as wiring and mechanical functions.

If the system goes down in an emergency situation, there will be emergency repair and labor costs, especially if this were to occur during the evening, on a weekend, or on a holiday when labor rates are higher. There will also be the labor cost for additional security and management staff to provide security coverage if the security system is not operating.

One method of reducing the cost to routine and emergency labor is to enter into an annual maintenance contract. It often will allow for a reduced rate for monthly or quarterly work on the security system, as well as emergency maintenance situations during the day, evenings, weekends, or holidays.

Replacement Cost

All things may pass, and that is true of security systems that become inoperable or antiquated. When designing and installing a new system, it is important to determine the life of the system. How long will it last before it needs to be replaced or becomes obsolete based on new hardware or software?

The manufacturer can most often advise on the life cycle of the system and potential future changes that may occur along with a time frame for such changes. Based on the life expectancy projection, a long-term budget should be established so that there are funds for the replacement of the security system at the anticipated replacement time.

The life cycle of the security system should also be a consideration when a system is first selected.

Cost-Benefit Analysis

When developing a security system, stakeholders must often prioritize requirements as part of the requirements engineering process. Not all aspects may be implemented due to lack of time, lack of resources, or changing or unclear project goals. It is important to define which requirements should be given priority over others.

Cost of Loss

Computing costs of a security system can be very difficult. A simple cost calculation can take into account the cost of repairing or replacing the security system. A more sophisticated cost calculation can consider the cost of having a security system out of service, added training, additional procedures resulting from a loss, the cost to a company's reputation, and clients. For most purposes, you do not need to assign an exact value to each possible risk. Normally, assigning a cost range to each item is sufficient.

One method to analyze the cost is to assign these costs based on a scale of loss as follows:

- Nonavailability of security system over a short term (7–10 days)
- Nonavailability of security system over a medium term (1–2 weeks)
- Nonavailability of security system over a long term (more than 2 weeks)
- Permanent loss or destruction of the security system
- Accidental partial loss or damage of the security system
- Deliberate partial loss or damage of the security system
- Unauthorized disclosure within the organization
- Replacement or recovery cost of the security system

Cost of Prevention

This includes calculating the cost of preventing each type of loss. This could include the cost to recover from (1) fire, (2) power failure, or (3) terrorist incident. Costs need to be amortized over the expected lifetime of the security system.

Return on Investment

In all areas of management to include the development of a total security system, return on investment (ROI) is a critical step in selling the system to top management and obtaining funding for the project.

A security investment such as a physical security system can enhance the security picture and improve the financial picture of the organization. Many security professionals have the technical security knowledge to sell a security system but lack the ability to show how security improvements can contribute to a company's profitability.

When making the business case for a total security system investment that will include software or hardware, it is imperative to capture the costs and benefits accurately and to present the results in compelling financial terms. Knowledge on how to quantify the security investment and the projected return in ways that top management and other financial stakeholders are used to seeing can be critical to obtaining endorsement of the security system.

Return on investment is a concept used to maximize profit to an organization for monies spent. It is used to determine the security system's financial worth. Return on investment is the annual rate of return on an investment. Developing a security system can be complex. Return on investment can be measured using two basic criteria: costs and benefits. The object is to establish a credible return on investment and also to define a high-value security system project by the benefits that it provides.

It is important to identify the purpose of the security system project. Is the project worth it? Will it improve security? If you can answer yes, the next decision is the priority of the security system project in the scheme of the total organizational goals. While security risk needs to be a priority, financial factors are a reality. Fiscally responsible planning and prioritizing will weigh in the project's favor in the decision-making process. The description of the project's purpose should include a clear statement of the need for the security system.

It is important to capture all the relevant costs of a project as it relates to return on investment.

Total cost of ownership (TCO) is the cost to an organization to acquire, support, and maintain the security system. TCO can be articulated in the following way:

TCO = cost to buy + cost to install + cost to operate + cost to maintain

Cost Factors

There are common cost factors associated with the development of a security system. It is vital to estimate both the extent and the timing of costs to be incurred during the security systems project. Typical cost factors for security systems may include the following:

- Security cameras/video
  - Cameras
  - Encoders
  - Fiber transceivers
  - Monitors
  - VCR, DVR, NVR
  - Mass storage
- Access control
  - Panels
  - Doors and locks
  - Readers
  - Gates
  - Other security sensors
- Communications
  - Leased line costs
  - Cost associated with interoperability of systems
- Cabling and power supplies
- Personnel associated with the security system
  - Receptionist
  - Credentialing
  - Contractor administration
  - Lock and key management
- Monitoring and control rooms
  - Alarm and video monitoring personnel
  - Operations support personnel
  - Physical security information management systems
  - Awareness and response systems
- General system-related costs
  - Engineering and design
  - Infrastructure and maintenance
  - Software and licensing
  - System deployment
  - Application integration
  - Administration and troubleshooting
  - User training

To be successful in selling the project, you must identify the benefits as they relate to return on investment. Start with the direct benefits, which are verifiable and easy to understand. Indirect benefits can be selectively included later if they contribute to the return on investment.

The return on investment can be justified based on the direct benefits attributable to the security system project. Direct benefits may include:

- Space improvements
- Wiring and communications infrastructure improvements
- Servers, applications, or systems improvements
- Storage increase
- Integration of systems such as security, fire protection, access control
- System maintenance and upgrades
- Training improvements

Indirect benefits are not easily measured. Productivity improvement is an example of an indirect benefit. Since indirect benefits may involve some subjectivity, separating indirect and direct benefits makes proposal evaluation easier, increasing its chances of receiving thorough consideration. IP-based physical security can be used to increase efficiency and provide labor reduction. The following list is an example of indirect benefits:

- Visitor management administration and control
- Segregation of duties
- Parking permit administration
- Property passes administration
- Employee time keeping
- System troubleshooting and maintenance
- Alarm correlation and response
- Emergency communication and notification

Once the cost and benefit data are collected, they must be analyzed to determine the return on investment. This can be accomplished by using the TCO comparisons as shown before. Presenting the return on investment should be done in a clear and concise executive summary. Presenting the return on investment of the proposed system in this manner can lead to the approval of the security system.

Capturing this advantage in quantifiable and credible terms will permit the calculation of return on investment. In the current business climate, it is crucial to justify the expenditure of a security system even if the risk analysis shows that it is vital to the security of the organization. Using the return on investment model will demonstrate the security director's business acumen and sensitivity to resource limitations. It will build the security manager's credibility with top management.

Determining the total security systems cost is an important aspect of the development of the proposed security system. It aids in the approval of the security system and the funding for the security system project by top management.

A complete analysis of the total cost will include system design cost, system installation cost, system operational cost, maintenance cost, IT cost, replacement cost, and return on investment. Following these guidelines in the development of the security system will yield results more favorable in the approval and funding of the total security system for the facility.[3]

References

1. Salamasick, M., and C. Le Grand. 2003. *PC management best practices: A study of the total cost of ownership, risk security and audit.* New York: Institute of Internal Auditors Research Foundation.
2. Fischer, R. J., E. Halibozek, and G. Green. 2008. *Introduction to security*, New York: Elsevier.
3. ASIS International. 2010. *Protection of asset manual.* Arlington, VA: ASIS International.

Chapter 18

Security Master Plan

Timothy Giles

Security Master Plan Strategy

An important aspect of the security master plan development process is to make sure that security strategies are linked to the strategies of the business, so that you can ensure that the program is moving forward in unison with the business. By doing this you will demonstrate to executive management that the security operation is no longer just a business expense, but rather an integral part of the business and contributes to the success of the business.

It is important to understand that although security professionals are focused on the many diverse risks that face our businesses and people, the executives who manage that business are not. They have many issues that occupy their time and thoughts on a daily basis. That is not to say that they do not care about these issues; they absolutely do. In fact, I have never met an executive who was not extremely concerned about any issue that might affect the employees or the business. I simply wish to point out that they are not as involved in them as we are. This process is the vehicle that will provide you the opportunity to bring these issues to the management team's attention through a business process and give you the platform for gaining the support the security function needs to manage effectively the risks that confront the business or institution.

Building a security master plan differs considerably from just conducting a site security assessment because you will not only need to identify the good and bad of the current programs, but will also need to help develop the corrective actions and long-term strategies. This would normally require that the person working on this master plan process have extensive knowledge and experience in all aspects of security programs and technology. However, this chapter provides the necessary guidance and information to help compensate for a lack of experience or knowledge and assists you to develop the plan.

The process defined in this chapter is designed to be utilized by an outside professional, a security consultant, as opposed to being performed by someone who works within the current security organization. However, it can also be performed by an internal professional, but in my opinion, you will find that with some areas of the process it will be difficult for an internal person to be completely objective. Areas such as defining the current skills and knowledge of the security organization will be especially difficult for him or her. Also, although I sometimes implement this process on my own, you have the option of supplementing your skills with others who may be more skilled in certain areas than you are. I find this team approach to be an effective way to achieve the end result.

Engaging the Stakeholders

It will also be important to put together a group of functional representatives from across the business to provide advice on where they believe there are currently areas that need change or improvements and how they perceive the recommended changes affecting the day-to-day operations of the business. Typically, these representatives would be from the following groups: facilities or engineering, human resources, information technology, manufacturing, research and development, and administration, as appropriate to the specific client. If the business has union workers, you may want to have a union representative in this group as well. The exact makeup of the group will depend on the business or institution that is being evaluated.

This group, referred to as "stakeholders," is the representative of all of the internal and possibly some external organizations that would be affected by changes to the security technology, policies, and practices. By involving this group in the process from the beginning, you will gain cross-functional support for implementing the necessary changes that will come out of the process. Of course, you may also encounter some resistance to some of the recommendations for change, but this will give the chief security officer (CSO) or director of security or you the opportunity to address these issues early on. Even if they are not fully resolved, you will at least have knowledge of what issues need to be addressed with the executives when it is time to meet with them.

I would add that in the corporate world it is commonplace today for many functions to hire outside consultants to do assessments of their operations and provide an unbiased view of what should be changed or improved. This is almost the standard with some functions, such as the finance and the information technology (IT) organizations. It is interesting to note that while there has been some change in recent years, typically the security community does not take advantage of this kind of independent review nearly as much as the other functions do.

I believe this is a change whose time has come because, as a community, we need to draw on the skills and knowledge of the experts within our profession more effectively and more consistently than ever before. As someone who has been a security director, I understand how difficult it is to manage just the day-to-day operations of your business and how little time there is to keep abreast of the fast-paced changes that engulf our industry. By having a consultant come in to look at the operation with a new set of eyes, you can gain immeasurable insight into what changes you should be focused on.

Although many of today's chief security officers or directors of security have a good insight into the technological changes that affect the security world and have their own ideas as to what direction they believe their business will take relative to these technologies, I have found that only some of them have actually documented this direction in a sound business plan and shared it with their management.

For example, many of the CSOs or directors of security that I have dealt with over the years who were utilizing magnetic stripe badges had never talked to their management team about migrating to proximity badges until they were in the process of requesting the monies to implement that change. In today's security world I believe you would not find many organizations that have a documented migration plan to move from proximity badges to utilizing either smart card or biometric (or both) technologies for their badges, just as you would not find many of them that have a documented plan to implement intelligent closed-circuit television (CCTV) software for their camera systems. However, I think if you asked the CSOs or directors of security, you would find that all of them believe they will move in these directions within the next few years. The security master plan process will provide them with the right vehicle to correct this situation.

What Should Your Security Philosophies Be?

This area is to be reviewed by the security consultant; however, the development of the philosophies should be done by the in-house security organization. If there is no in-house security organization, then the consultant should attempt to work with the in-house person who manages the security contract to develop the appropriate philosophies to follow. First, I believe that the philosophies of the security organization should reflect the culture of the overall business. Next, they should reflect the leader of the security organization's business beliefs and, to some extent, personal beliefs and character. These philosophies are the basis upon which the security program is built. For example, some of the beliefs that I have used include the following:

- "Respect for the individual." This respect should be for each and every individual, including the ones who are believed to be violating your security policies and procedures.
- "Excellent service to the customer." This applies to both internal and external customers and at every level of the security organization.
- "Excellence as a way of life." Every action should always be done to the best of one's ability.
- "Managers and supervisors must lead by example." This is a critical aspect of projecting how all employees should act. "Do as I say, not as I do" will never work.
- "We should always be a good corporate citizen." For the security organization, this is reflected in the way you deal with and support the many public organizations you interface with, such as law enforcement, fire departments, and rescue services.

Of course, these are only examples of some of the philosophies I have used. This is truly a personal choice for the person who is in charge of the security organization. It is doubtful that you will encounter many CSOs or directors of security who have actually written their philosophies down and shared them with their staff. I firmly believe it is an exercise worth undertaking and that it can be a guide for the entire security organization.

In many cases the company will have written philosophies or principles that it publishes for all employees. If this is the case, then I would recommend to the CSOs or directors of security that they expand on those to help the security organization understand how it should be reflected in the day-to-day operation by the security staff, and they should also add some of their own philosophies to them and in support of them.

If the organization utilizes contract security officers, it is very important that they are also made aware of the organization's philosophies. It may be necessary that they or the contract manager translate these into statements that reflect how these philosophies actually affect the day-to-day duties of the officers as well. This is usually done through the post orders; however, they may need to be elaborated on to get the desired result.

Contract Security Relationship

It is exceedingly important for the organization to have a "partner" type of relationship with the contract security force. This can be a delicate situation because the client does not want them to believe they are "employees" of the organization, but should want them to see themselves as an integral part of the team. This is typically achieved by making sure the chain of command is always used when dealing with the security force. It is also critical that their own management, both on-site and off-site, have discussions with them on occasion about maintaining the right relationship with the "client." It will be very important for you, the security professional, to determine if this relationship is sound and appropriate.

A common development in this environment is that you will see one of the lead people of the contract force begin to develop a personal relationship with some of the lead people on the in-house security or other staff. Over time this can manifest itself into problems where they begin acting as if they are employees of the organization, instead of the contract force. Likewise, the organization begins to treat them more like employees and even gives them more power in the relationship than they should have. Whenever this situation develops, the only effective way I have found to correct it is for that person to be taken off of that site.

What Should Your Security Strategies Be?

Before you begin the process of defining or redefining the security organization's strategies, you must first gain an understanding of the strategies of the business. You do this by interviewing the appropriate executives of the company: the CFO, COO, and so on. For the next 5 years, you need to know:

- What growth do they anticipate?

- Do they expect any product or service changes?
- Is the expansion or reduction limited to the existing facilities or will new ones be added?
- Do they expect any overseas expansions or mergers?
- Are there any major layoffs or outsourcing activities planned?

Some of this information will be considered to be highly confidential, especially any mergers or layoff activity, but you need to understand these directional moves if you are to plan how to deal with them from a security standpoint. It is not necessary for you to know all of the details; for example, you do not need to know whom they plan to merge with or whom they plan to outsource work to. However, you will need to know what countries are involved if your client will have any stake or ownership in the relationship.

If the person performing this master plan activity is an outside consultant, the executives may prefer to share this information only with the in-house director of security or chief security officer. If there is no in-house staff, the consultant will need to discover as much of this information as possible and may need to sign a confidential disclosure agreement (CDA). (I believe a CDA should always be part of the contract with the consultant.)

The security organization's strategies deal with all aspects of the program from policies and procedures to technology and staffing. These strategies should be documented so that they reflect where the organization is now and where it is going. You have probably heard this before, but I believe strongly in the saying, "If you don't know where you are going, you won't like where you are when you arrive!"

In order to implement new security strategies, CSOs or directors of security should first address the process of change. This is an area where you, the security professional, can provide advice and counsel, but implementation must be performed by someone in house. It has been my experience over the years that most people are afraid of change. They would prefer that everything just stay as it is. So the question the CSOs should be asking of themselves is this: "Is change a friend or foe?" The answer to this question is really quite simple: "It is up to them!" Change is a topic that is discussed continuously in the business world. But, as the adage says, "Talk is cheap!"

As an example of implementing change I would cite the most dramatic project that I have undertaken in my career. If you have not been involved in a major change effort, then perhaps my experience can help you to understand the complexities of this effort. As a part of the reengineering effort in IBM, we reorganized the internal security operation in September 1994. We took the security professionals who were managed site by site by nonsecurity personnel and brought them into one single structure, managed by security professionals. However, this did not in and of itself make change happen. What it did do was to provide the opportunity for constructive, consistent, and rapid change.

Over the next 2 years we reduced costs by approximately 30%, we increased customer satisfaction to 94%, and we significantly increased our own security employees' morale. In September 1997, I was awarded the security director of the year recognition by *Access Control & System Integration* magazine. As people passed on their congratulations to me, I explained that I take credit for one thing primarily, and that is creating the environment where "change" is a "friendly" activity. The accomplishments of our organization are directly attributed to our own people embracing the concept of change and making it happen.

So exactly what did we do to create this environment? Basically, we did three things. First, we implemented the use of project teams on as many different aspects of our security business as we could think of. These teams had two goals to accomplish: find the best internal or external practice for the specific area they were looking at and—even more importantly—increase open communications across the organization.

Second, we implemented a measurement program to find the defects in our processes. To make this successful, I declared this to be a "no fault" measurement program. The primary "failure" in this program would be if we did not find problems. The secondary failure would be if we did not fix the problem.

Third, we launched a massive campaign to do national contracts and centralized systems to eliminate as many redundancies and inefficiencies as possible. All of this combined translated into massive change for our people and our strategies in the way we implemented security.

We knew that the only way we could be successful was for our people to see this as something that would be good for them—each and every one of them—personally. To make this happen we first had to convince them that change was absolutely necessary to the survival of IBM and our jobs. You might think this would be obvious to all of us considering our company's financial performance over the early 1990s, but some people have a way of convincing themselves that they are not part of the problem. Therefore, what we had to do was to convince them that change had to happen and we had two choices:

1. Deny the need, resist the change, and FAIL.
2. Embrace the need to change and DRIVE that change.

If we, the security professionals, truly and fully accepted this, we had the power to decide our future! If we did not drive change in our organization, someone else would and we would have much less control over the outcome.

One of the primary tools that we provided to our project teams to do their analysis was the implementation of an internal benchmarking program followed up with a detailed resource and task analysis program. After implementing many of the changes and realizing the benefits of those changes, we then launched an external benchmarking effort. These data demonstrated that we were significantly more cost competitive than any of the other companies we compared with.

As any good business manager can tell you, the best resources of any company are its employees. I believe that this group of security professionals is the best of the best, but I acknowledge that I might be slightly biased on this point; however, the proof is in the results! It is important to remember that change is not something that you do and it is finished. Instead, it is an ongoing process that must be continually driven from senior management down through the organization and by the employees up through the company. This is why it is essential that you create the right environment for change to flourish.

A critical part of that environment is your own attitude! Your employees will know very quickly if you are just giving "lip service" to this process or if you are serious. Just as the scenery changes as you travel down a road, your business and even you and your employees must be in a continuum of change. If you are, you will not just succeed, but you will also have ongoing success! It is this environment that makes it very important that you have documented, long-term strategies and that you reevaluate those strategies on a regular basis. After all, that is the map you will be using for your trip.

So, what are your clients' or company's strategies? As I said earlier, they should cover all aspects of their programs. It would be very difficult for me to suggest any generic strategies because there are many variations depending on the business they are in. As you develop them, you should utilize the functional team, the "stakeholders" that I spoke about earlier, to assist. Here are some examples of the areas that should be addressed:

- Policies
- Education and awareness programs
- Badge wearing.
- Clean-desk policy
- Visitor and contractor controls
- Employee involvement and responsibilities
- When and how to have armed off-duty police officers on-site
- Investigations
- Use of hidden cameras along with determining who should be involved in the decision to use them
- Use of a polygraph for interrogations
- Whether or not to prosecute employees or others when a crime has been committed (even a minor crime)
- Technology
- What technologies might be utilized in the future and when, where, and why
- The migration plan for moving to the new technologies
- The anticipated end of life of the current technologies in use
- Developing a replacement schedule for existing equipment
- Staffing
- The use of armed or unarmed security officers documented with the reasoning for the decision
- Which positions can or cannot be contracted, regardless of whether they currently are or are not contracted
- What style of uniforms should be worn and why

As you go through the process in documenting their strategies, they will find that they are already following several strategic lines; they just may not have documented all of them before. A good example of this is the use of unarmed security officers. I do not like to have armed security people onsite except in rare applications such as a nuclear plant or a top-secret installation. Obviously, many CSOs or directors of security feel the same way because the majority of businesses in the United States use unarmed officers. However:

- How many of these security managers or businesses have documented that decision to demonstrate that it was a well-conceived strategic decision?
- Was executive management involved in or at least apprised of the reasoning for this decision?

- If a workplace violence shooting were to occur on-site, would they be prepared to defend in court their decision of unarmed officers?

Having these strategies well documented can be invaluable in situations of litigation or even when a decision about an unusual situation has to be made in a timely manner. Their documented strategies should always be their guide.

Technology Migration Strategy

I would also like to discuss the issue of "migration strategies" for their changes in technology. If you believe they might be moving to a different technology for access control, for example, it is very important that they have investigated the issues around migrating from the current technology to the new one. If the client has a single site or even just two or three sites, the migration can be relatively easy to accomplish; nevertheless, it still requires a detailed plan, which includes having test locations and education for the end users. By the way, I have seen situations where the end users were not properly educated in the use of the new technology and this set back the conversion by several months; the security team spent countless hours struggling to convince the end users that the new technology was the right solution for the business.

However, if the client has a large number of sites, there needs to be a plan that addresses how the client will operate during the migration to the new technology. For instance, when we looked at migrating IBM from magnetic stripe access control cards to proximity cards, there was no existing solution that allowed us to have both technologies in use at the same time without actually mounting both types of card readers side by side so that employees could gain access regardless of which card they were carrying.

The solution offered by our vendor was to take out the old technology and put in the new one. That might be a workable solution for someone who has only a few locations, but for a company that has hundreds of locations and employees who need to be able to access multiple sites, that is not an acceptable remedy. To resolve the problem we developed our own approach. We went to several vendors and asked them to develop dual-technology cards and card readers. Of course, they wanted us to fund the development work for these new products, but we convinced them that this was an investment they needed to make not just for us but also to assist any large company that needed a solid migration path to the newer proximity technology. Eventually they agreed, and the new products began to hit the market.

Although this provided the hardware solution to our dilemma, that was only part of the final solution. Issues such as importing or exporting databases and conversions of data, education of users, and determining who needed dual-technology cards and who did not, along with numerous other minor issues, all had to be researched and resolved prior to the start of the migration. When you are changing technologies for hundreds of thousands of end users and hundreds of locations, you also have to have a detailed timing plan as well. You cannot make that kind of a change in a few weeks. My message to you and your client is this:

- Do not assume that the vendors have the right plan for migration.
- Do not let yourself be limited by what currently exists, especially if it does not solve your problem.
- Spend some time investigating others who have made the changes that you are considering and learn from their experiences.
- Make sure you budget some additional funds to help educate the end users on how to use the new technology.
- If the current vendor they are using has never migrated a client of their size to the technology they are considering, they should investigate other vendors to find one that has.

One other approach to be considered is to introduce the new technology slowly into their business. If they currently have proximity access control and they want to move to biotechnology, I would recommend introducing biotech into their high-security areas first. They might want to try the different biotech readers to see which they like best, so they might use the palm print reader in one area, the fingerprint or iris scan reader in another, and so on. This will give them the opportunity to gain experience with the new technology and get feedback from the users and management.

If and when they decide to implement it on a larger scale, it is no longer something that is brand new to the end users, as they will all have heard about its use and the migration can proceed in a much smoother progression. Additionally, end users typically feel much better about new technologies if they have been able to provide input into the decision.

Equipment Replacement Schedules

I will also review the subject of developing a replacement schedule for existing equipment. The best way to run a security operation is to develop a list of every piece of equipment that the security department owns. This list should be detailed to include the following information:

- Name
- Model number
- Serial number
- Date purchased
- Supplier
- Purchase price
- Location installed
- Supplemental information (For example, if it is a camera, you should also include the lens specifications here. If it is a radio, how many channels are on it, etc.?)
- Manufacturer's recommended life cycle
- Projected replacement date

With this information you can establish a replacement schedule for the equipment similar to what the facilities engineering department does for the equipment that supports the building. Once you have this piece of documentation, it can be used during the budget cycle to project future expenditures for keeping the equipment and systems in peak operating condition. Another use of these data is with the contract vendor that is maintaining the systems. Instead of the client budgeting for replacing the equipment, they could have the vendor build the replacement schedule into the annual contract for maintaining the systems. Some businesses find that to be a more acceptable approach to this issue.

One other consideration relative to the equipment that is installed is the documentation of the location and wiring specifications. The wiring specifications should include both the communication cables and the power system. Most facilities these days have their drawings on CAD/CAM or other computer software files. These are files that can be accessed by computer that show every aspect of the facility. However, I frequently find that the security systems have not been included in these drawings; they typically only include the base building information that is used by the facilities organization.

This leads to numerous problems whenever these systems need to be upgraded or replaced. It also creates havoc with the systems when there are renovations performed at the facility because the contractor performing the renovations would not have knowledge of the security systems. As a security professional you should review a sample of the drawings and determine if they are fully documented. If they are not, that should be part of your recommendations, and getting them documented should be a part of the security master plan action items.

Chapter 19

Security Foresight

Paul R. Baker

Introduction

In 1938, Douglas "Wrong Way" Corrigan set a course to fly from New York to California, but ended up landing in Ireland—because he seemingly misread his compass. Knowing your destination and reading your compass correctly are your best approaches to the future. The world of security is changing faster then we sometimes realize or wish to acknowledge. In addition to the age-old problems of violence and crime, security executives now contend with international terrorism, environmental damage, energy disruptions, and potential pandemics. To these protracted and almost universal problems one can add the prospect of unexpected and often violent natural events like earthquakes, floods, or hurricanes.

There will certainly be less time to plan and control responses to fast moving and unpredictable events. In the future there will be ever greater penalties and burdens for failing to react effectively. Dwindling reaction time is a feature of the modern international scene. If we are to get ahead of the response curve, then we must increasingly look ahead and anticipate. This means increasing our planning and preparation well before any potential event. As a result, we need to develop more imaginative and dedicated threat assessments, as well as a more structured approach to measuring the likely pitfalls and warning signs.

What is foresight? According to Richard Slaughter, foresight is common sense. Experience (interaction of memory and foresight, identity and purpose) is read upon yet-undetermined situations to (1) avoid dangers, (2) reduce risks, and (3) manifest highest potential.[1]

Strategic Foresight

Strategic foresight is the ability to create and maintain a high-quality, coherent, and functional forward view and to use the insights arising in useful organizational ways. It is used to detect adverse conditions, guide policy, shape strategy, and explore new markets, products, and services. It represents a fusion of future methods with those of strategic management.[1]

Strategic foresight should sound familiar to security professionals: avoid danger—reduce risk. Security organizations revolve around two tasks: (1) risk and vulnerability assessments (identifying threats) and (2) continuity planning (preparing for disruption).

A ship's captain would never leave port without navigational tools, maps, global positioning systems, and a crew that was seasoned. The same goes for security professionals who need the tools to initiate risk assessments and contingency planning.

In a strategic planning process, there are four fundamental questions:

Where are you now?
Where are you going?
Where do you want to be?
How are you going to get there?

Visual ways to address these types of questions help the mind to "see." Seeing can help identify issues and opportunities, organize information, prioritize, clarify thinking, and set goals on a personal and/or organizational level.

Visual tools and techniques are the most effective when they are set in the right framework. One of the keys to good visioning is good framing of issues. The combination of questioning and visual techniques can bring out the creative thinker in even the most task-oriented person.

Are you focused on the right questions? Indicators are pointers of potential change over time, whether that is in the short or long term. If they are applied to a standard risk profile, it is possible to convey a sense of direction to conventional threat assessments. Combining the historical data of previous incidents with potential indicators provides a trend path for a range of threats. This approach can apply as much to forecasting hurricane strikes in the Carolinas as forecasting robbery incidents in a large city. Such an exercise is far more meaningful than supplying a single snapshot in a rapidly changing world.

The viewpoint increasingly applied to these considerations is that an organization is an "open system"; its growth and survival are dependent on the nature of the environment it faces.[2] If any organization is subject to its environment, its future is tied in with factors (economic, political, technological, environmental, and sociological) that will inevitably come into play as change takes place within that environment. This is important because, where opportunities exist, it will be apparent in those future changes.

> Change has become the essence of management, so to survive and prosper in the future, you and your organization will have to perfect "outside-in" thinking skills: to relate information about developments in the external world to what is going on internally and this will provide for significant issues which may emerge from unexpected places.[3]

Foresight Techniques

From a security professional standpoint, utilizing techniques that provide risk assessment for your organization and that develop security foresight is a proactive approach toward security leadership. There is a process for bringing to the table others within the security organization who can provide ideas and support the overall security requirements of the operation.

Begin with bringing a group of security professionals together and present the group with three specific tools and techniques in order to perform a sound futures mapping of the security industry. These consist of mind mapping, environmental scanning, and future mapping.

Mind Mapping

Mind mapping is a powerful technique that can help to develop a strategy or expand thinking on a subject. The "map" uses words, lines, logic, colors, images, and links to draw out associations and stimulate thinking. The technique works as well in large group brainstorming sessions as it does one on one with a coach.

While there are many different mind-mapping systems, the basic process involves expanding on ideas using key words. The objective is to make a complex or thorny topic easier to understand, explore, or remember.

For an example of mind mapping, have the security professional create a simple mind map dealing with closed-circuit television (CCTV):

1. Draw a circle in the middle of a blank sheet of paper and write a project, goal, dream, or idea in the center of the circle.
2. Draw lines (spokes or branches) radiating out from the central circle.
3. Write down thoughts/ideas that relate to the central circle at the end of each spoke and circle them.

Here is an example of how to mind-map CCTV. These ideas are concrete and in the present. They focus on what is needed today and how it can be accomplished. As security professionals, we are aware of the need for CCTV and the protection it affords an organization. This is one step to prepare for the future and the consistent need for risk assessments and evaluations. After putting CCTV in the center of a circle, these ideas might come to mind:

- Recording devices. Do I use a digital recording device or back up all images on a hard drive or do I still use tape drives? Am I just recording or actively watching?
- Human factor. Do I have someone actively watching the CCTV? How long will someone be posted in front of the cameras? What is the ergonomic design of the monitoring station?
- Number of cameras. How many cameras do I need? How many will be active/passive? Are they activated by motion or alarm action?
- Surveillance versus monitoring. What am I looking for? Do I need to identify personnel or do I need to just oversee an area?
- Cost. What is the cost of the camera system I want to employ? IP technology, digital, or analog?

Environmental Scanning

A second proactive tool for the security organization is environmental scanning. This is a process of scanning the horizon, to see what might lie ahead, and considering how that might impact an organization's critical assumptions. Similar to a lookout on a ship's mast, the scanner enables an organization to spot issues, increase response times, and proactively take advantage of them when they occur.[4]

Scanning comes first, but it is only a means to an end: The real value comes in when the organization looks beyond the issues it has identified to visualize the implications for the particular industry or profession.

The amount and variety of data suggest the need for systematic procedures to collect, organize, and integrate information relating to relevant macroeconomic, social, political, and technological indicators for the product markets in which a company is involved. This should be consolidated with internal information into a usable form to strengthen the market strategy.[5]

Here is a preface to an environmental scanning report. Note how the report takes into account the critical decision issue facing the security organization and links that it is transactional and macroenvironmental:

Profession's Work/Industry Environment

Security issues have clearly moved up on the corporate agenda, but what remains unclear is who owns this ever expanding and more prominent security domain. It consists of many disparate elements, including physical security, cyber security, risk management, safety, and disaster management/business continuity. The security environment requires a copious degree of cooperation. In this field, security will be the ultimate director in providing protection from a multitude of potential threats. The security personnel at the RAND Corporation are conscientious and professional in their daily duties.

MACROENVIRONMENT

The ramifications of September 11 have woven security into all of our daily lives. Various degrees of increased physical security measures have been in effect throughout the organization since the terrorist attacks, and will continue for the foreseeable future. This is the new world in which we live and it requires an ever vigilant security program in order to protect the organization's most important asset, its people.

FOCUS OF THIS SCANNING REPORT

From a security standpoint the question and the strategic foresight will be: What are the potential hazards and risks associated with working in the government research and development arena? What are the required areas that security needs to address in the future? This scan will identify risks that need to be addressed and potential threats that are on the horizon.

Future Mapping

Scenario planning tools have been around for decades and are useful to help anticipate change, predict the elements of different scenarios and develop strategies to be able to shape each possible future.

Today there are many models that take scenario planning to the next level. Dr. Canton's *future mapping* tool is an excellent way to map an emerging industry or problem. It makes the distinction between forecasting (getting advance information about the future based on analysis of existing conditions and trends) and foresight (the ability to see what is emerging):

> Maps are so much more revealing, though, when they're studied closely in the context of when and where they were created. Maps are snapshots of time that chronicle change, the evolution of thinking about navigation, conquest, commerce, threat and opportunity. They are more than just geography. Maps change when new information is introduced.[6]

This tool creates scenarios based on key change drivers, trends, and "forces that can shape the future of an enterprise, market, industry, society or civilization."[7]

In business as in map making, strategic thinking about the future requires constant redefining, change, updating, and refinement. You are always dealing with new information and change. You need to change your map constantly—to navigate the waters of change.

Also, since at any given time our worldview is distorted, at best incomplete, the idea of process to map the future makes a lot of sense. By better visualizing the future, you can make more informed choices about business or strategies toward shaping a better future. Future maps are visual tools that can be used to think strategically about the future and to make sure you wind up where you want to go.

Consider what a future map would look like for the future of security. One of the major drivers for security is technology. Try to identify several technological advances that security may see in the next 15 years. This will keep you looking forward at new areas that security needs to address. Continuous strategic thinking in a changing world is vital if security is to maintain a winning position. Anticipate change rather than simply react to it, and adapt its strategy as necessary to keep moving ahead.[8]

Here are five features that might fill out a future map for technical advances in security:

- **Biometrics.** Retinal scanning, fingerprints, and voice comparisons will be required before entry is granted into secured areas.
- **Body scanning.** Equipment will sniff out hazards, toxins, and explosives on people and in packages.
- **Smart cards.** Genomic personal identity cards will be required to enter or leave areas. They will have unique DNA markers for each individual.
- **Integrated cameras.** Cameras working with access control and intrusion detection systems will grant entry through facial recognition and will know when to alert guards of potential alarm activity. By 2015, the United States will have more than ten million closed-circuit television cameras—more than half operated by government agencies and the rest by private security and corporations. The average citizen in New York City will be photographed 500 times a day.[7]
- **GPS control.** Every chip in every product will be online, linked to GPS satellites that can track every human on the planet.

As security professionals we are tasked with protecting people, information, and property. It requires a vigilant approach to anticipate what might be potential threats in the environment.

Aiming to make the organization more sensing, agile, and prepared provides a clearer purpose, direction, and context for security leadership. Looking beyond security to resiliency may provide the change in perspective that organizations need to balance security and risk management with the organization's strategic foresight. In the next 5 to 10 years there will be an ever changing world. The security of an organization's information and people will be just as important as it is today.

Cultivating strategic foresight requires cooperation with other security professionals and law enforcement in order to anticipate security trends. Communication, cooperation, and a well-developed continuity plan will provide for a security program that is well prepared for future events before they occur. Even the Bible makes reference to the need to be prepared by looking down the road at potential risks. "Be sure of this: if the master of the house had known the hour of night when the thief was coming, he would have stayed awake and not let his house be broken into."[9]

In general, a horizon scanning and forecasting function is evaluated by managers and executives who have used it as a positive tool, along with the recognition of the need to consider longer planning horizons and differing organization structure to incorporate a forward planning function. Corporate-based respondents strongly prefer that such forecasts be conducted by and be part of the ongoing corporate planning function, while governments utilizing strategic planning prefer to have it allocated to a separate department.[10]

A security operation would use this type of strategy to (1) identify potential threats; (2) stay alert to political and economic trends, opportunities, and/or threats; (3) allocate resources to areas in need of increased security; (4) develop action plans and scenarios for potential threats; (5) learn about where the threat may come from (inside or outside); and (6) evaluate the risk.

The tool of mental models could also work well, particularly alongside strategic planning efforts. Strategic foresight is based on the principle of planning from the future back to the present, not the typical strategic planning approach of working from the present toward the future. The use of mental models helps change the way the organization thinks.[4]

Chris Argyris says that everyone has "theories of action," which are a set of rules that we use for our own behaviors as well as to understand the behaviors of others. However, people do not usually follow their stated action theories. The way they really behave can be called their "theory in use." It is usual to:

1. Remain in unilateral control
2. Maximize winning and minimize losing
3. Suppress negative feelings
4. Be as rational as possible—by which people mean defining clear objectives and evaluating their behavior in terms of whether or not they have achieved them[11]

Revolutionary change in mental models cannot occur without a radical change in thinking. Radical changes in thinking ultimately require the use of analogy and metaphor. "Ideas about organization are always based on implicit images or metaphors that persuade us to see, understand, and manage situations in a particular way...The challenge facing modern managers is to become accomplished in the art of using metaphor to find new ways of seeing, understanding, and shaping their actions."[12] A similar personal discovery was made by MIT Media Lab guru Michael Schrage:

> I had thought of myself as someone who was very good at—and loved to play with—innovative ideas. Give me a clever metaphor, model, simulation, or prototype and I was a cat with a ball of yarn. There was no idea that couldn't be transformed or unraveled simply by batting it around with the right mix of rigor, curiosity, and playfulness...I was forced to confront some simple truths about what I was doing: I really wasn't playing with ideas; I was playing with *representations* of ideas.[13]

Thomas Edison and his team of "muckers" working at the Menlo Park lab would continually glimpse the future through the use of analogy and metaphor.

> Throughout his career, he drew directly or by analogy upon his repertoire as he engaged simultaneously in a rush of many diverse electromechanical and electrochemical developments. In this case, as so often in Edison's or other people's work, a principle, property, or device, developed in one context, and ultimately inapplicable in the context, fit beautifully as a solution in an entirely different one.[14]

How far into the future did metaphor and analogy allow Edison to see? "On New Year's Day, 1871, more than three decades before the Wright Brothers' historic flight, Edison speculated that an engine can be so constructed of steel and with hollow magnets...and combined with a suitable air propelling apparatus wings...as to produce a flying machine of extreme lightness and tremendous power."[15]

Edison's foresight led to a "virtuous cycle" of learning and design, and continual shifting of mental models. Consider the development of the phonograph:

> A single page from Edison's notebooks beautifully captures his remarkable facility for mixing and matching concepts. It shows three designs for recording sound, from the time Edison first displayed his phonograph. Those pictures foretell the main directions the recording industry would take throughout the first half of the twentieth century. One sketch, illustrating the design Edison went

on to market commercially, shows a stylus pressed against a cylinder resembling a rolling pin. The so-called "cylinder phonograph" derived directly from a cylinder version of his recording telegraph. A second drawing features a grooved disk not unlike an LP record. It sprang from the basic version of his telegraph recorder; the device that led to the discovery that sounds could be captured on paper or foil. The third drawing foreshadows the tape recorder, with paper tape running under a stylus. The project scholars believe that Edison got this idea from his work on earlier printing and chemical telegraph systems, which had similar configurations... When Edison was thinking about developing moving pictures, in a patent caveat he announced, "I am experimenting upon an instrument which does for the eye what the phonograph does for the ear." He went on to describe the parallel between the spiral of images that make up what we now call film and the spiral grooves on the phonograph record.[15]

Another tool is "wildcard scenarios," which allow an organization to further examine possible futures and to think strongly about how these futures might affect the organization's strategic planning—what it has set out to do.

Wild cards are axes for providing an operation with the avenue to allow its imagination to run wild. They put the other scenarios into perspective and make them more plausible.

One of the most important characteristics of any organism is its ability to adapt to new and changing environments:

> Flexibility of perspective is critical. You simultaneously focus on questions open for the unexpected. Like a hunter, alerted to the presence of prey by the snap of a broken twig, you learn to pick out a key piece of vital information in the dizzying flood of words, images, sounds and numbers that most of us swim in.[16]

It is impossible to know what the future will hold for us, but we need to be prepared and one way of doing this is through the use of scenarios. "Using scenarios is rehearsing the future. You run through the simulated events as if you were already living them. You train yourself to recognize which drama is unfolding. That helps you avoid unpleasant surprises, and know how to act."[16]

Uncertainty about the future has encouraged the use of wild cards in strategic foresight. Wild cards

- Have the ability to produce early warning and are useful for general inspiration
- Advance brainstorming and stimulate unconventional ideas
- Increase creativity, inventiveness, and fresh thinking
- Produce new ideas, products, or policies

However, the disadvantages of wild cards include (a) significant dependence on the observer and (b) the difficulty in validating the process.

Strategic planning involves distinct thought processes that are analytical and convergent. In this instance, planning is considered single-loop learning in that thinking and acting are within a certain set of assumptions and potential action alternatives.[17] Examples of single-loop learning include cost cutting, de-layering, or reengineering. These are actions that organizations have used in the past to arrest deteriorating performance. Comparatively, strategic thinking and foresight emphasize synthesis and creativity and can be characterized as double-loop learning that involves challenging existing assumptions and alternatives. Double-loop learning facilitates foresight activities and creative alternatives such as innovating new services to expand markets or forming alliances to complete. The argument is that *strategic foresight leads to different responses from what was done in the past*.

Most organizations have a short-sighted approach to doing business. Short-term profits and immediate concerns to deal with competitors outweigh spending time on developing foresight into the future in the form of strategic thinking or strategic planning. The *urgent* drives out the important and the future is left largely unexplored. For the most part, senior executives devote less than 3% of their time to building a corporate perspective on the future. In some companies, it is less than 1%. In order to maintain a competitive edge, senior managers must be willing to devote considerably more time to planning for the future.[18]

Organizational transformation must be driven by a point of view about the future of the industry: How do we want this industry to be shaped in 5 or 10 years? Industry foresight is based on insights into trends in technology, demographics, regulations, and lifestyles that are used to develop new industry rules and create a new competitive environment. Industry foresight is a synthesis of many people's visions.

Change is inevitable, and the rate of change appears to be increasing as time goes on. The issue for managers is whether that change will occur unexpectedly in a crisis atmosphere or with the benefit of foresight in a calm and considered manner.

References

1. Slaughter, R. 1999. *Futures for the third millennium*. St. Leonards, NSW, Australia: Prospect Media.
2. King, W., and V. Narayanam. 2000. Environmental scanning and forecasting in strategic planning—The state of the art. *Long Range Planning* 33:130.
3. Ashley, W., and J. Morrison. 1997. Anticipatory management tools. *Futurist* 31 (5):47.
4. Marsh, N., M. McAllum, and D. Purcell. 2002. *Strategic foresight: The power of standing in the future*. Melbourne, Australia: Crown Content.
5. Douglas, S., and S. Craig. 2005. *International marketing research*. Englewood Cliffs, NJ: Prentice Hall.
6. Canton, J. 2006. *The extreme future*. New York: Dutton Publishing.
7. Ringland, G. 2006. *Scenario planning: Managing for the future*. New York: John Wiley & Sons.
8. Hines, A. 2006. Strategic foresight: The state of the art. *Futurist* 40 (5):18–21.
9. Holy Bible. The New American Version, Matt. 24:43. Wichita, KS: Catholic Bible Publishing, 1994.
10. Schwartz, P. 1991. *The art of the long view*. New York: Doubleday.
11. Argyris, C. 1991. Teaching smart people how to learn. *Harvard Business Review* 4 (2).
12. Morgan, G. 1997. *Imaginization*. Thousand Oaks, CA: Sage.
13. Schrage, M. 2000. *Serious play*. Boston: Harvard Business School Press.
14. Jenkins, R. 1985. A record for invention: Thomas Edison and his papers. *IEEE Transactions on Education* 27 (4):191–196.
15. McAuliffe, K. 1995. The undiscovered world of Thomas Edison. *Atlantic Monthly* December: 80–89.
16. Schwartz, P. 1991. *The art of the long view*. New York: Currency Doubleday.
17. Heracleous, L. 1998. Strategic thinking or strategic planning. *Long Range Planning* 3 (3):481–487.
18. Hamel, G., and C. K. Prahalad. 1994. Competing for the future. *Harvard Business Review* 72:4.

Chapter 20

Security Leadership

Paul R. Baker

Introduction

Leadership is dispersed throughout all segments of the society—government, business, organized labor, professions, minority communities, universities, social agencies, and so on. Most leadership today is an attempt to accomplish a purpose through large, intricately organized systems. Individuals in all segments and at all levels must be prepared to exercise leadership, taking initiatives and responsibility and using their knowledge to solve problems at their level.[1]

What Is Leadership?

Leadership is a complex phenomenon involving the leader, the follower, and the situation (Figure 20.1). Some leadership researchers have focused on the personalities, physical traits, or behaviors of leaders; still others have studied the relationship between leaders and followers. I will attempt to demonstrate the traits needed to be a successful security leader.

Others have defined leadership as follows:[2]

- The creative and directive force of morale
- Directing and coordinating work of group members
- An interpersonal relation in which others comply because they want to—not because they have to

Figure 20.1 Leadership is an abstract concept that, in a fast-moving, complex world, entails a number of factors.

- The process of influencing an organized group toward accomplishing its goals

Although many such definitions may seem confusing, it is important to understand that there is no single correct definition. The various definitions can help us appreciate the multitude of factors that affect leadership, as well as different perspectives from which to view it.

For a security leader these definitions and ideals present the foundation to take initial theories and make them into a workable situation. Leadership is the foundation from which a security organization can project itself into a world-class operation.

Purpose of Leadership

Leadership is a way of focusing and motivating a group to enable them to achieve their aims. It also involves being accountable and responsible to the group as a whole.[3] A security leader should (1) provide continuity and momentum and (2) be flexible in allowing changes of direction.

Ideally, a security leader should be a few steps ahead of his or her team, but not too far from the team to be able to understand and follow. Communication and direction are key components in providing the purpose for the security group.

Effective Leadership

For a security professional to become an effective leader, some areas will require him or her to search from within. In this respect, the following is a list of the necessary requirements for becoming an effective leader:

- Effective leaders make others feel important. If goals and decisions are self-centered, followers will lose their enthusiasm quickly. Leaders emphasize others' strengths and contributions, not their own.[4]
- Leaders promote a vision. Followers need a clear idea of where leaders are leading them, and they need to understand why that goal is valuable to them.[5]
- Leaders follow the golden rule. They treat the followers the way they enjoy being treated. An abusive leader attracts few loyal followers.[4]
- Leaders feel free to admit mistakes.[5]
- Leaders may criticize others in private, while praising them openly in public. Public praise encourages others to excel, but public criticism only embarrasses and alienates people.
- Leaders stay close to the action. They need to be visible to the members of the organization. They talk to people, visit other offices and work sites, and ask questions. They observe how business is being handled. Often, they will gain new insights about their own work and find new opportunities for motivating their followers.
- Leaders often make a game of competition. The competitive drive can be a valuable tool if they use it correctly. They set team goals and reward members who meet or exceed them. Leaders examine their own failures and celebrate groups' successes.[6]

Leadership is just one of the many assets a successful security professional must possess. The main aim of a security leader is to maximize the output of the organization through protective implementation. To achieve this, the security leader must undertake the functions of organizing, planning, staffing, directing, and controlling in the organization. Leadership is just one important component of the directing function. A security professional cannot just be a leader; he or she also needs formal authority to be effective. "For any quality initiative to take hold, senior management must be involved and act as a role model."[7]

Security leaders are also people who are able to express themselves fully. The key to full expression is understanding oneself, knowing one's strengths and weaknesses, knowing how to maximize one's strengths, and working on one's weaknesses in order to overcome them. People who can express themselves fully know what they want, why they want it, how to communicate what they want with others, and how to achieve their goals.[4] The key to understanding is learning from one's own life and experience.

It is not easy to become a leader, but it is a much easier to learn how to lead than most of us think because everybody has the capacity for leadership.[8] Many scholars define leadership as beauty—"It is hard to define but you will know it when you see it."[9] Leadership is like any human talent: Some people are more inclined toward it than others, but there are some techniques that can be taught. Leadership occurs when people with the talent and training find themselves in circumstances that enable them to put that talent and training into action.

Why Security Leaders Are Important

Security leaders are important for several reasons. First, they are responsible for the protection of organizations. The success or failure of all organizations rests on the protection of assets. Second, the change and upheaval of the past years show that we need anchors in our lives that can guide us. Leaders fill that need. Third, there is a pervasive, national concern about the integrity of our institutions. It is not enough for a leader to do things right; "he must do the right thing."[10] Furthermore, a security leader without some vision of where he or she wants to take the organization will lack a sense of camaraderie among the troops and a comfort level with upper management.

Understanding the Basics

According to Alan Bryman,[3] leaders seem to share some, if not all, of the following characteristics. This also holds true for security leaders:

- ***A guiding vision.*** A leader has to have a clear vision of what he or she wants to do and must be strong enough to stick to this vision when faced with numerous obstacles or failures.
- ***Passion.*** A leader loves what he or she does. A leader who communicates passion gives hope and inspiration to other people.

Integrity. Three essential parts of integrity are self-knowledge, candor, and maturity. A leader knows his or her strengths and weaknesses and potential. A leader is true to himself or herself and knows what he or she wants to do and why. In this way, the leader is able to invent himself or herself, grow, and succeed. Candor is the key to self-knowledge. It is based in honesty of thought and action. It is an unchanging devotion to one's principles. A leader does not compromise on principles to please others. The third essential part of integrity is maturity. A leader should take learning seriously, as well as be truthful. By exhibiting these qualities, a leader can encourage them in others.

Trust. Trust is a quality that cannot be acquired but must be earned. It is more of a product of leadership than an ingredient.

Curiosity and daring. The leader wonders about everything, wants to learn as much as he or she can, and is willing to take risks, experiment, and try new things. The leader does not fear failures because he or she will learn from them.

Achievement. Leaders have a relatively high desire for achievement. High achievers obtain satisfaction from successfully completing challenging tasks, attending to standards of excellence, and developing better ways of doing things.

Ambition. Leaders are very ambitious about their work and careers and have a desire to get ahead. Effective leaders are more ambitious than nonleaders.

Tenacity. Leaders are better at overcoming obstacles. They have the "capacity to work with distant objects in view and have a degree of strength of will or perseverance."[10] Leaders must be tirelessly persistent in their activities.

Initiative. Effective leaders are proactive. They make choices and take actions that lead to change instead of just reacting to events or waiting for events to happen.

Energy. Leaders should have a high level of energy. Leaders have been characterized as "electric, vigorous, active, and full of life."[4]

Leaders continually expand their competence—their ability to do things. Effective leaders see "life as a mission not as a career." Their nurturing sources have armed and prepared them for service. They radiate positive energy. This positive energy is like an energy field or an aura that surrounds them and that similarly charges or changes a weaker, negative energy field around them. They also attract and magnify smaller positive energy fields. When they come in contact with a strong negative energy source, they tend either to neutralize or to sidestep this negative energy. Sometimes they simply leave it, walking away from its poisonous orbit.

A leader believes in the unseen potential of all people. A leader does not feel built up when he or she discovers the weaknesses of others. Leaders are active socially, having many friends and a few confidants. They think in terms of continuums, priorities, and hierarchies. Their actions and attitudes are proportionate to the situation: balanced, temperate, moderate, and wise. They view life as an adventure. Leaders regularly exercise the four dimensions of the human personality: physical, mental, emotional, and spiritual. They participate in some kind of balanced, moderate regular aerobic exercise. They exercise their minds through reading, creative problem solving, writing, and visualizing.[4]

Emotionally, leaders make an effort to be patient, to listen to others and show genuine empathy, to show unconditional love, and to accept responsibility for their own lives, decision, and reactions. Spiritually, they focus on prayer, scripture study, meditation, and fasting.

Are Leaders Born or Made?

Are leaders born or made? This question continues to dominate the study of leadership today. Volumes of research have been written; however, little or no conclusive evidence has been established. The topic of leadership remains elusive. One difficulty in discussing the topic is ascertaining the definition. Burt Nanus and Warren Bennis report some 350 definitions of "leadership."[11]

After very extensive research on leadership, many scholars have concluded that, in the majority of cases, genetics and early family experiences play a significant role in developing the personality and character needs that motivate the individual to lead. Other research studies have indicated that the origins of leadership go beyond genes and family to other sources. Work experiences, hardship, opportunity, education, role models, and mentors all go together to craft a leader. Current research suggests that experiences on the job play an important catalytic role in unlocking leader behavior. There seems to be no substitute for learning through doing, making mistakes, and improving with time.[12]

One survey of 200 executives at highly successful companies concluded that early in their careers leaders had opportunities to lead, to take risks, and to learn from their successes and failures.[13] Survey results specifically identified the following as important developmental opportunities:

- Challenging assignments early in a career
- Visible leadership role models who were either very good or very bad
- Assignments that broadened knowledge and experience
- Task force assignments
- Mentoring or coaching from senior executives
- Attendance at meetings outside a person's core responsibility
- Special development jobs
- Special projects
- Formal training programs

Thus, leadership must still be understood as a complex equation of birth and early childhood factors, shaped by later life experiences and opportunities.

Good Leadership

> The leader who is poor of spirit knows that his employees are intelligent people who, many times, know more of the details of the job and thus good leaders understand the concept of team. They are only as good as the part or the employee under them. They do not consider themselves an expert in the field but rather a continual learner who utilizes the knowledge of their subordinates to brainstorm ideas. The total team effort will bring forth ideas from all members and will allow the subordinates to gain confidence in their leadership skills and problem-solving abilities, [and] have worthwhile advice to give. This is a key premise of total quality leadership—to teach the employees how to solve problems, develop solutions, and then trust them to do the work.[14]

Setting the example is an important leadership skill. If you want someone to learn something new, you must show him or her. If a leader does not practice what he preaches, he lessens the effectiveness of the message. Setting the example will institute a respect for the leader and provide the subordinates with the same vision and goal that the leader is attempting to achieve. "The team needs to know how to get the job done, and they lack know-how and often the courage to get started. The inspiring figure of a Sergeant actually taking the time to show his troops how to do it means much."[15] It is also an effective way to show others the proper way to conduct themselves.

I was fortunate to find a company that encompasses a lot of what the ideal organization should be to its employees. The MITRE Company is on the Fortune 500 "Top Companies to Work for" list. The company believes in a happy and productive workforce and provides its employees with a Wellness Department that includes a staffed fitness center, nurse's station, cafeteria, coffee kiosk, and learning center. It is truly a joy to belong to a community rather than a business. We even called the company "Mother MITRE" because the company will take care of you and nurture you for your entire career.

This is a good Catch-22 situation, in which a good environment produces good leadership and vice versa. This was emphasized at a recent staff meeting in which my boss asked all managers to generate new ideas and thoughts concerning security. He wanted a collaborative effort from managers and employees on how to improve security measures at the company. He was showing that he was a humble leader who does not hover over his employees, but rather has respect for his subordinates by asking for their input on security issues that affect all members of the company. "A humble leader shows respect to all, whether they are superiors or subordinates, because the leader who is poor in spirit recognizes that many people know more than he or she does and, as such, shows respect to everyone."[16]

Bad Leadership

Regardless of how much emphasis the business world places on leadership skills, there will always be bad leaders. Some leaders look at a problem and assume the fault lies with the employees, rather than with them. Some leaders find themselves falling into the sin of pride.

> You can't go around with your chest stuck out and your head full of yourself with the belief you are all knowing and all seeing, because there's always somebody around who is more than willing to challenge your authority, or worse, to undermine it. A good leader does not put much emphasis on the title, role, or trappings of leadership, but rather, concentrates on the responsibilities of leadership and keeping up good relations with the group.[17]

I was a Maryland state trooper for the majority of my working career and it was considered a paramilitary style police department. Everything was very rigid and regulated. I was accustomed to it, coming directly out of the Marine Corps, but it lacked the camaraderie that the military brings—the taking care of one another. The one example that sticks in my mind is the day a new barrack commander arrived and he gave his standard boilerplate speech "If you're right, I'll be behind you 100%."

I began to think about that statement and it started to anger me; if I'm right, I expect to be supported. I want to know what support I will have if I'm wrong. If I was putting my life on the line, I felt that I deserved just as much support whether I was right or wrong. On the road, when split-second decisions are being made and there is not the luxury of sitting behind a desk and reviewing the incident several days later, I want a commander who understands mistakes are made and who will back me no matter what the outcome.

Being held accountable for wrong decisions is a fear for some commanders. "Making mistakes only helps you become a better person, manager, etc. I use the analogy of a basketball player that has no fouls. If they are not going for the ball and taking chances with their opponent, then they are trying hard enough. Take a chance and don't be scared."[18] In this environment, where a leader refuses to support his or her troops, the leader breeds fear and inhibits creativity, innovation, and often productivity. "Employees who know that the leader has their interests at heart are willing to commit themselves to corporate tasks."[19]

Like most police departments there is a ranking system—private, sergeant, and so on, in which as you obtain rank you receive more responsibility and authority. Within the Maryland State Police it was a very autocratic, top-down system of leadership. Only commanders had ideas; no one without substantial rank could possibly have an idea. And the majority of the higher-ranking ideas fell along the lines of "this is the way I did it when I was on the road" or "this is the way I was taught in the police academy." It was sacrilege to ask for advice from the people who were actually on the streets doing the job. The brass's mantra was always: "That's why they are paying me the big bucks—to make decisions." This is the reason the headquarters name became known as the "Crystal Palace." Troopers obtain rank through a functions test, based on the patrol and investigative manuals. There are no educational requirements or leadership training; if you score high enough on the test, you are promoted.

Another example of a bad situation was when I worked for a security manager that was completely out of her league. This was a manager who had no experience dealing with a security operation, but for some unknown reason had been placed in charge. Any time security matters arose and decisions had to be made, it was the classic "deer in the headlights" look. When things went wrong, it was always someone else's fault.

According to Barbara Kellerman, "Leaders are incompetent for various reasons. Some lack experience, education, or expertise. Others lack drive, energy, or the ability to focus. Still others are not clever enough, flexible enough, stable enough, emotional[ly] intelligent enough."[20] This was the situation I was confronted with and many of the "enough" as outlined by Kellerman were "tattooed all over my boss."

It did not take me long to realize that my supervisor was overmatched for her position. I initiated a complete security survey for the facility and found some very significant deficiencies in security; some were in violation of the NISPOM (National Industrial Security Programs Operations Manual—the security regulations that all government contractors are required to support). When presented with evidence of the most serious security lapses, she stated, "But we have passed the inspections." My response was that, yes, we had passed the inspections and possibly the inspector missed the violations, but it was the responsibility of the government contractor to provide protection of classified information and the facility that stores the classified information. I made it clear that she was the FSO (facility security officer) and it was her responsibility to ensure that the NISPOM was being followed, and if information concerning problems with security were brought to her attention she needed to address them. Instead, she decided that when the inspector found the violation, then she would act, but in the meantime—why worry?

Not only did I find my supervisor incompetent, but in some regards she was completely negligent in her position as the FSO.

My position was to be a buffer between my direct reports and her. If I could take the brunt of her incompetence and continue with the day-to-day operations, then this would be a success in the short frame of things. It is imperative that subordinates know that you will provide a buffer for them; in turn, they will give the security operation their very best. In order to turn that philosophy into action, the people working for us must know that we will go to bat for them when they make honest mistakes. If they do not have that confidence in us, then we cannot expect them to stick their necks out.

What can you do to protect yourself and survive working for an "incompetent" leader? According to Carole Nicolaides in "Dealing with Incompetent Leaders,"[21] some quick tips include:

1. **Do not make it a personal matter.** This is a hard one, simply because working for an incompetent boss is such a personal matter. Remember, that most of these leaders do not have a problem directly with you, but they too are frustrated and are shouting loud their own insecurities—most likely mirroring to you things that they should be doing.
2. **Observe your boss.** It might sound funny, but notice what is going on around your boss. In case you've known or worked with your boss before and you observe a sudden change, then your next step should be to take action right away. The problem could be as simple as someone asking him something way out of his league, or someone talking to him about you and your team. Whatever the reason might be you need to act and confront your boss as soon as possible. If you do this at the beginning, you might be able to stop a snowball effect—not only for you but also for the entire team. Confrontation does not come easy for most people, yet if you seek a constructive conversation, have an open mind, avoid turning it into a personal attack, you might be able to ease tensions with your boss and also improve his position.
3. **Accumulate facts.** Nothing is irrelevant if you work in an unhealthy environment. You need to make sure that you accumulate all the things that matter for your career—the good as well as the bad stuff. Good things that you've done, bad things that have happened to you, and things that you could have done better. The key here is to have nothing against you, nothing that will give people permission to talk about you and question your character.
4. **Know your value.** You might feel beaten down, overworked, underappreciated, and doubtful about your true value. Grow up! Things happen and your value does not diminish simply because one cannot see your true value. If you are a professional, do a good job, and the people that work with you will see a direct contribution to the team's success. Then be sure that you have created your own evangelists—people who will tell others about your true value.
5. **Expand your network.** Now, more than ever, you need to think that working for a large company is not very different than working on your own. You need to learn to promote

yourself. People need to know who you are, within your company and outside your company. Successful business owners never stop networking. There are so many things you can learn simply by networking. The key here is to find 2 or 3 networking initiatives that you feel comfortable doing and commit to them.

6. **Seek for comfort outside your office.** Many people often make this mistake. They work for an incompetent boss and they start complaining about her or him to a "good friend" who also works for the company. For whatever reason this might happen because you are seeking comfort or love. Sometimes you simply need a sounding board in order to release the penned-up stress. Do it outside the office and avoid discussing your problems with others with whom you work.

Going from Bad to Good

Throughout my career I have come across good and bad leaders. The good ones all seem to care about people, inspire confidence, mentor their subordinates, maintain a professional demeanor, have a vision for the team, listen, and be dependable. The bad managers are demeaning, autocratic, impatient, inefficient, lacking in vision, demanding, incompetent, and without the experience or skills to manage teams.

I do not believe that anyone intends to be a bad leader. I do feel that a bad leader believes he or she is doing an excellent job and is focused on the business at hand. I also believe that a bad leader will not change unless it is in his or her best interest to do so. "Bad leaders will not change unless they calculate the benefits of good leadership as greater than the benefits of bad leadership."[22]

Leaders need to look in the mirror occasionally and do a self-evaluation. I always sit back and ponder: "Am I doing the best I can to motivate and encourage my employees? Are they excited to come to work each day?" I have been fortunate that I have a management structure that encourages good leadership values and is interested in the employees' well-being.

Kellerman lists several ways a leader can correct and strengthen his or her capacity to be effective and ethical:[23]

- *Limit your tenure.* When leaders remain in a position of power for too long, they tend to acquire bad habits.

- *Share power.* When power is centralized, it is likely to be misused, and that puts a premium on delegation and collaboration.
- *Don't believe your own hype.* There is always room for improvement; don't believe that you walk on water—that is the kiss of death.
- *Get real, and stay real.* Don't lose touch from where you came from.
- *Compensate for your weaknesses.* Surround yourself with expertise in areas that you are not strong in and listen to their recommendations.
- *Stay balanced.* Don't be a workaholic; don't be more dedicated to the job than to your family and friends.
- *Remember the mission.* If it is security, stay focused on protecting people, information, and property.
- *Stay healthy.* Physically, emotionally, spiritually, and psychologically
- *Develop a personal support system.* Back to the concept of you can't do it all by yourself. You need a team effort and the support that arises when all are working toward the same goal.
- *Be creative.* The past should never determine the future nor narrow the available options.
- *Know and control your appetites.* These include the hunger for power, money, success, and sex.
- *Be reflective.* Emphasis on self-knowledge, self-control, and good habits.

Smile and be polite to your subordinates or at least maintain a calm demeanor. This is easy to say and hard to do at times. If you can maintain a consistently pleasant and calm disposition it will make working for you a more satisfying experience for your people. If you are having a bad day, try to keep it to yourself. There will be times when the situation is so grim that if you have a smile on your face people will think you are insane. Even when things are not going well for you or your security group, it can only make it worse by being sullen or by ranting and raving. When the boss is upset, everyone else gets worried and starts looking for some cover.

Do not expect your subordinates to have the same level of security leadership understanding as you do:

> Sometimes common sense is an uncommon virtue. Don't assume wisdom on the part of your subordinates. Wisdom comes from life experience; very few of us are born with an abundance of it. Your subordinates most likely have not been around as long as you, and if they have, they probably haven't had the same leadership experiences you have. If they do something unwise—before you hammer them—take a hard look and ponder if you might have done something similar at their age or experience level.[24]

Good and bad people can equally hold the elusive qualities of leadership. George Washington demonstrated a quality of leadership that embodies the highest aspirations and ideals of the qualities desirable in a leader, and people like Gaddafi, Stalin, and Napoleon demonstrated the depths of inhumane cruelty that leaders can sink to, and yet still retain the power of leadership. "Employees want leaders who are honest, open, and who keep the organization moving in a positive direction during both calm and stormy seas."[25]

As security leaders we are entrusted with enormous influence on the lives of individuals and our leadership plans are only good if they include a way to keep the well-being of the employee (rather than ourselves) in prominence.

> Times have changed and even though it might seem hard to work for someone that you know is not suitable for his or her position, remember things and people appear to us to teach something. The sad reality is most "incompetent" leaders do not get fired; they just move on and reinvent themselves in new companies. The chance that you will either work with the same leaders or someone like them again before your career ends is great. However, if you manage to stay calm and think about the lessons you've learned and how to counteract incompetent behavior, you will have all the wisdom needed in order to become a better leader yourself in future jobs."[21]

Forging the Future

To master the competitive environment, the security leader must first understand the challenges of the new millennium. Understanding is the first step. The security leaders of the future will be those who take the next step to change the culture.

According to Warren Bennis,[26] there are 10 personal and organizational factors for the future:

Leaders manage the dream. All leaders have the capacity to create a compelling vision, one that takes people to a new place and then to translate that vision into reality. Managing the dream can be broken down into five parts. The first part is communicating the vision. The other basic parts of managing the dream are recruiting meticulously, rewarding, retraining, and reorganization.

Leaders embrace errors. Leaders are not afraid to make mistakes and admit them when they do. They create an atmosphere in which risk taking is encouraged. For them failure is not a crime.

Leaders encourage reflective backtalk. Each leader knows the importance of having someone in his life that would tell them the truth. The trusted person is reflective because it allows the leader to learn, to find out more about themselves.

Leaders encourage dissent. Leaders need people around them who have contrary views, who are devil's advocates— "variance sensors" who can tell them the difference between what is expected and what is really going on.

Leaders possess the Nobel factor. Optimism, faith, and hope. Optimism and hope provide choices. The leader's worldview is always contagious.

Leaders understand the Pygmalion effect in management. Leaders expect the best of the people around them. They know that the people around them change and grow. If you expect great things, your associates will give them to you. At the same time, leaders are realistic about expectations. Their motto is to stretch, not strain.

Leaders have the Gretzky factor, a certain touch. Wayne Gretzky, the best hockey player of his generation, said; "It's not important to know where the puck is now as to know where it will be." Leaders have that sense of where the culture is going to be, where the organization must be if it is to grow. If they don't have it as they start, they do when they arrive.

Leaders see the long view. They have patience. They have a long-term perspective that is rewarding.

Leaders understand stakeholder symmetry. Leaders know that they must balance the competing claims of all the groups with a stake in the corporation.

Leaders create strategic alliances and partnerships. They see the world globally, and they know it is no longer possible to hide. The shrewd leaders of the future are going to recognize the significance of creating alliances with other organizations whose fates are correlated with their own. This is how the leaders forge the future.

New Definitions for Leaders in the Twenty-First Century

Rob Rogalski, corporate security director for the RAND Corporation, defines the new era of leadership as: "Leadership is no longer the exclusive domain of top management. Everyone in a security operation has an obligation to lead" (personal communication, 2011). In today's competitive business world, organizations need visionary leaders, facilitative leaders, inspiring leaders, and collaborative leadership. Leaders of all types are arising at every level of an enterprise. Leadership is no longer exclusively top-down, but also bottom-up.[27]

A second important quality leaders will need is that they must facilitate excellence in others; trust, respect, listening, inspiration, educating, and recognizing creativity can do this. The third important quality leadership will need is a humanistic aspect of leadership. In order to lead a balanced organization effectively, leaders will need to cultivate many different dimensions within themselves. This may call for being committed and hard driving and simultaneously having a healthy, peaceful mind and a fit body. The last essential skill for twenty-first century leaders is the mastery of change. To prepare for the next century, many changes will include difficult adjustments in both the structure and size of organizations. Leaders will need to guide their organizations through these disruptions and yet retain a motivated, committed organization that can build for the future.

In a recent interview, Robert Crandall, chairman and president of American Airlines, stated:

> I think the ideal leader for the twenty-first century will be one who creates an environment that encourages everyone in the organization to stretch their capabilities and achieve a shared vision, who gives people the confidence to run farther and faster than they

ever have before, and who establishes the conditions for people to be more productive, more innovative, more creative and feel more in charge of their own lives then they ever dreamed possible.[28]

Based upon current research, the next generation of successful leaders will have certain things in common. The most important ones are a broad education, boundless curiosity, boundless enthusiasm, belief in people and teamwork, willingness to take risks, devotion to long-term growth rather than short-term profit, commitment to excellence, readiness, virtue, and vision.[29]

Strategic leadership is the process that an organization uses to assess its current position in the industry, develop its vision for the future, and design the steps necessary to achieve that vision. In order for a company to develop its strategic plan, it must accomplish a series of steps, including gathering and analyzing information about the company and its environment, identifying critical issues facing the organization or the industry in general, developing a strategic vision, reviewing and revising the mission as the vision is clarified, and, finally, developing strategic goals and strategies. The ultimate goals of the strategic planning process are to assess the current status of the organization and to make decisions about the future of the organization based on that assessment.[30]

What is always most critical is the level of involvement of people that have been developing the strategic plan. In general, if people are involved in the process in meaningful ways, they will be ready to do their part in implementing the plan that results from the process. And, of course, the flip side is also true: If people whose involvement is critical to the successful implementation of the plan have not been involved in meaningful ways, they will not be excited about the plan and will not be very motivated to take some responsibility for implementation.[30]

Overall leadership is the constant blood flow for a security professional. Without the continual flow of leadership moving within your essence, there is little for an organization to accomplish and even less for followers to buy in to. For a security leader, the development of leadership traits will provide for a substance that will generate trust and a feeling of purpose all the way down to the guard on the midnight shift. There is no direct course to becoming a security leader; however, it is essential to continue toward the goal. This has been an outline of potential areas of need for a security leader and the necessity to prevail as a leader within a security operation. If one leadership style does not feel comfortable, there are others that have been identified to try. Just make sure, as a security professional, that you keep moving in a direction of that affords you the ability to lead a security organization.

> Leadership is not having a magnetic personality…It is not "making friends and influencing people." Leadership is lifting a person's vision to higher sights, the raising of a person's performance to a higher standard, the building of a personality beyond its normal limitations.
> —Peter F. Drucker

References

1. Lord, R. G., and K. J. Maher. 1991. Leadership and Information processing. *Journal of Management* March: 80–95.
2. Morris, S., and R. Stogdill. 1989. Different definitions and conceptions of leadership. *American Journal of Sociology* 56:149–155.
3. Bryman, A. 1996. Leadership in organizations. In *Handbook of organizational studies,* ed. S. R. Clegg and C. Hardy, 200–210. Thousand Oaks, CA: Sage Publications.
4. Covey, S. 1992. *Principle-centered leadership,* 209–230. New York: Simon & Schuster.
5. Stogdill, R. 1990. *Bass & Stogdill's handbook of leadership. Theory, research, and managerial applications,* 415–610. New York: The Free Press.
6. Smith, H. L., and L. M. Kruger. 1993. A brief summary of literature on leadership. *Administrative Science Quarterly* 16:321–323.
7. Predpall, D. 1994. Developing quality improvement processes in consulting engineering firms. *Journal of Management in Engineering* May/June: 30–31.
8. Bass, B. 1985. Leadership and performance beyond expectation. *Harvard Business Review* 7:40–47.
9. Bennis, W. 1979. Leadership theory and administrative behavior. *Administrative Science Quarterly* 4:110–117.
10. Bennis, W. 1995. *On becoming a leader: Handbook of leadership,* 304–309.
11. Bennis, W. G., and Nanus, B. 1985. *Leaders: The strategies for taking charge.* San Francisco: Harper-Collins.
12. Pfeffer, J. The ambiguity of leadership. *Academy of Management Review* 2:100–110.
13. Lord, R. G., C. L. de Vader, and G. M. Alliger, Meta-analysis of relation between personality traits and leadership perception: An application of validity generalization procedures. *Journal of Applied Psychology* 71:402–410.
14. Winston, B. 2002. *Be a leader for God's sake,* 24. Virginia Beach, VA: Regent University.
15. Miller, C. *The empowered leader: 10 Keys to servant leadership,* 112. Nashville, TN: Broadman and Holman Publishers.

16. Winston, Op. cit., 24-25.
17. Loeb, M., and S. Kindel. 1999. *Leadership for dummies,* 292. New York: Wiley Publishing.
18. Ortiz, C. *40+: Overtime under poor leadership.* Bloomington, IN: Author House.
19. Winston, Op. cit., 29.
20. Kellerman, B. 2004. *Bad leadership: What it is, how it happens, why it matters.* Boston: Harvard Business School Press.
21. Nicolaides, C. Dealing with incompetent leaders http://www.careerknowhow.com/guidance/incompetent.htm
22. Kellerman, Op. cit., 232.
23. Ibid., 233.
24. Miles, R. 2003. *Organizational strategy, structure, and process.* Stanford, CA: Stanford University Press.
25. Winston, Op cit., 9.
26. Bennis, 1995. Op. cit., 405–415.
27. McFarland, L., and J. Childress. 1993. *Twenty-first century leadership: Dialogues with 100 top leaders.* New York: The Leadership Press.
28. Wren, J. T. 1995. *The leaders' companion.* New York: The Free Press, p. 456.
29. Bennis, W. 1990. Managing the dream: Leadership in the 21st century. Training. *Magazine of Human Resource Development* 27 (5):44–46.
30. Martinelli, F. 2002. Strategic planning. The Nonprofit Management Education Center. http://www.uwex.edu/li/learner/q-a1.htm

Index

A

Access control, 97–99, 114
    anti-passback, 99
    basic components, 98
    central station design, 221
    common intrusion tactics and countermeasures, 76–77t
    cost factors, 276
    data centers, 259–261
    design costs, 270
    gates, 88–89
    goal of entry control, 97
    guards' roles, 103–105, 202
    head end, 100–103
    ID badges, *See* Badges; Card readers
    integrated guard system, 99
    layers of defense, 17, *See also* Protection in depth
    mantraps, 105
    parking facilities, 50, 52, 97
    penetration testing, 70–75
    receptionist role, 81, 86, 101
    sensitive compartmented information facilities, 234–235
    system integration, 152
    turnstiles, 106–107
    violation monitoring, 75
    visitor management, 64, 101–103
Access roster, 2–3
Accountability, 313
Acoustical protection, 231–233, 245–246
Acoustic sensors, 113, 115–116
Administrative security, sensitive compartmented information facilities, 246
Advanced video code (AVC), 134
Adversary sequence diagram (ASD), 36–37
Aero-K, 267
Air contamination, 67–68
Alarm systems, *See also* Video surveillance
    access control violation monitoring, 75
    acoustic sensors, 113, 115–116
    alarm assessment, 223–224
    balanced magnetic switch, 114
    bank ambush features, 253
    false-alarm rates, 117
    fire detection/alerting, 263–265, *See also* Fire protection systems
    holdup alarms, 254–255
    interior intrusion detection, *See* Intrusion detection systems, interior
    layers of security, 18
    monitoring and responding, 103, 201, 219, *See also* Security control center
    night depositories, 251
    nuisance alarms, 105, 117, 223–224
    perimeter intrusion detection, *See* Intrusion detection systems, perimeter
    power, 236
    retrofitting, 56
    sensitive compartmented information facilities, 235–236
    silent, 18, 254–255
    vaults, 248
    windows, 112, 235
Alfred R. Murrah Federal Building, 82

All-clear notification, 65
Ambush alarm systems, 253
American Society for Industrial Security (ASIS), 35, 38, 221
Analog megapixel video, 141
Analog video cameras, 130–131, *See also* Cameras
Analog video surveillance, 123, 127, *See also* Video surveillance
Analogy and metaphor, 300–301
Anti-passback protection, 99
Application programming interface (API), 152
Architects and security design, 37
Armed versus unarmed guards, 213–214, 287–288
Assembly site, 65
Assets
  list, 8
  loss categories, 16
  target identification, 12
  value rating, 15
  vulnerability assessment, 13–16
Audit control, 75
Authentication factors, 175
Automated teller machines (ATMs), 249–250
Availability, integrity, and confidentiality, 17, 59

B

Badges, 2, 98–99, 175, *See also* Card readers
  anti-passback protection, 99
  garage entry, 50
  guards' role in monitoring, 104
  open area parking, 52
  penetration testing, 71, 72
  security personnel, 211
  technology migration, 281, 288–289
  for visitors, 101–103
Balanced magnetic switch (BMS), 114
Banks and financial institutions, 247
  ambush features, 253
  ATMs, 249–250
  bullet-resistant glass, 252
  dye packs, 18, 251–252
  guards, 255–256
  layers of security, 17–18
  location risk assessment, 255–256
  night depositories, 250–251
  pack tracking systems, 252
  safes, 248–249
  teller cash recyclers, 251
  vaults, 247–248
  video surveillance use cases, 124
  video systems, 253–254
Barbed wire, 87, 89
Barriers, 86–89
  bollards, 47
  fences, 87–88
  gates, 88–89
  walls, 89
Batons, 212, 213
Beam type smoke detectors, 264
Benchmarking, 286
Biometrics, 30, 175, 176–177
  attacks on and countermeasures, 194–197
  data center access, 260
  deploying, 186–189
    enrollment, 186–187
    template storage, 188–189
    verification process, 187–188
  facial recognition, 177–178
  fingerprint, 178–180, 195, 199
  future mapping, 299
  hand geometry, 180–181
  iris recognition, 181–182
  metrics, 190
    error trade-off curve, 194
    failure to acquire, 191
    failure to enroll, 191
    false rejection/false acceptance rates, 192–194
  multibiometric systems, 184–185
  palm print, 180
  privacy issues, 190
  retina scans, 198
  technology migration, 281
  vein pattern recognition, 182–183
Bistatic microwave sensor, 120
Body scanning, 299
Bollards, 47
Budgeting, *See also* Costs
  security plan adjustments, 23–24

total system cost determination, 269
    working with contractors, 39–40
Buffer zones, 31
    standoff distance, 30
Building construction projects, *See* Security construction projects
Building shape, 45
Bullet-resistant (BR) glass, 112, 252
Business continuity planning, 6, 68

C

Cameras, *See also* Closed-circuit television; Video surveillance
    analog video, 130–131
    ATMs, 250
    cost factors, 276
    day/night capability, 134–136
    design costs, 270
    HDTV, 126
    infrared (IR), 94
    integrated systems, 299
    IP video, 131
    IP video encoders, 132–133
    layers of security, 18
    lenses, 137
    license plate capture, 134
    motion-activated, 114–115
    network communication, 163–164
    pan-tilt-zoom (PTZ), 127, 131–132
    power over Ethernet (PoE), 128, 131, 136, 163
    progressive scan CCD, 137
    SD/SDHC memory, 132
    smart video, 121
Campus video surveillance, 124
Capacitive sensors, 179, 198
Carbon monoxide detectors, 118
Card readers, 28, 30, 75, 114, *See also* Badges
    access control system violations, 75
    anti-passback, 99
    biometric systems, 189
    data center entrance, 260
    design costs, 270
Cash recycling, 251

Central station design, *See also* Security control center
    alarm assessment, 223–224
    costs, 270
    design requirements, 221–223
    developing an operation, 219–221
    network operations center, 260
    secondary amenities, 223
Certifications, 215
Certified protection professional (CPP), 215
Chain-link fencing, 87, 89
Chain of command, 64, 283
Change, 284–288, 304
    resistance to, 5–6
Charged coupled device (CCD), 130
Chemical air contamination, 67–68
Chemical attack, protective design features, 31–32
Chief security officer (CSO), 280–281
Chroma mode capability, 135
Cipher lock, 167
Class A extinguishers, 265
Class B extinguishers, 265
Class C extinguishers, 265
Class D extinguishers, 266
Classified information security, *See* Sensitive compartmented information facilities
Class K extinguishers, 266
Cleaning crews, 260
Clear zone area, 48, 81
Closed-circuit television (CCTV), 27, 28, *See also* Cameras; Video surveillance
    alarm assessment, 223–224
    bank systems, 253–254
    effectiveness in areas with single exits, 82
    HDCCTV, 141–143
    illumination, 89, 93–94, 137–138
    integrated systems, 75, 101, 299
    intrusion detection systems, 121
    mind mapping, 296
    technology migration, 281
    traffic monitoring, 46
Closed storage, classified information, 237
Cloud computing, 144–148
    community cloud, 147–148

hosted video solutions, 145–147
hybrid (public/private) cloud, 148
private cloud, 147
public cloud, 147
Coaxial cable, 142
Coaxial strain-sensitive cable, 120
Code of conduct, 217
Color temperature, 138
Communal areas, 82
Communication plan, 70
Communications
    cost factors, 276
    locations for emergency systems, 52
    operations center system, 222–223
    planning issues, 27–28
    sensitive compartmented information facilities, 244–245
Community cloud, 147–148
Compression, 129, 133–134
Computer-aided drawing (CAD), 37, 290
Computer physical security, 262
    portable devices, 72, 234
Concertina, 87
Confidentiality, 17, 59
Construction permits, 271
Construction projects, *See* Security construction projects
Construction review, 40
Construction security plan, 227
Consultants, 280, 281, 284
Containers, 174, 236
Contaminated air, 67–68
Continuous lighting, 91
Contract central station, 113
Contractors
    budget issues, 39–40
    finding, 43
    requirements, 44
    review process, 40
    sensitive compartmented information facility construction, 237–238
    working with, 38–39
Contract security guard force, 208–209, 283
    awareness of security philosophies, 283
    hybrid (proprietary/contract) force, 209, 220
Control center, *See* Security control center

Convenience versus security, 4–5, 7
Cost-benefit analysis, 273, 277
Cost factors, 275–278
Costs
    change orders, 38
    cost-benefit analysis, 273, 277
    cost factors, 275–278
    IT-related, 272
    loss-related, 274
    maintenance, 273
    operational, 271–272
    of prevention, 274
    replacement, 273
    resource allocation for loss prevention/mitigation, 13
    return on investment, 274–275, 277
    security personnel, 207, 217, 272
    security plan adjustments, 23–24
    security plan implementation, 30
    system design, 270
    system installation, 271
    total cost of ownership, 275, 277
    total system cost determination, 269
Counseling, 217
Crime prevention through environmental design (CPTED), 44, 79–82
    commonsense approaches, 86
    comprehensive building security plan, 84–85
    organizational, mechanical, and natural design approach, 84
Critical building components, 50, 62

D

Data centers, 259–263
    access control, 259–260
    common security mistakes, 260–263
    fire protection, 263–265
    network operations center, 260
Day/night cameras, 134–136
Defense in depth, *See* Protection in depth
Defibrillator, 223
Delay, 19
Delta barriers, 47
Denial of service, 16, 130

Design and planning, *See* Security planning and design
Design cost, 270
Destruction loss category, 16
Detection, 19
Deter-detect-delay-respond, 19, 61, 247
Deterrence, 19
  technology capabilities, 28–29
Digital signal processing, 130
Digital video recorders (DVRs), 143–144, 222, 250
Disabilities, persons with, 65, 66
Disaster recovery planning, 68–69
Disciplinary action, 217–218
Disclosure loss category, 16
Dispatch center, *See* Security control center
Door-closing device, 109
Doors, 109
  evacuation planning and procedures, 64
  glass, 109
  layers of security, 18
  lighting, 89
  locks, 110–111
  perimeter, 109
  ready to exit (REX), 32
  request-to-exit (REX) systems, 111
  security control room, 222
  sensitive compartmented information facilities, 228, 235, 236, 242–243
Double-loop learning, 303
Drills, 64
Dual-technology sensors, 117–118
Dumpster diving, 74
Duress alarms, 255
Dye packs, 18, 251–252

E

Earthquakes, 22
EASI model, 36
"Easy Pass," 97
Eavesdropping, 72
Edison, Thomas, 300–301
Electrical power costs, 272
Electrical system, 230
Electric locks, 110
Electric strikes, 110

Electromagnetic radiation (EMR), 233–234
Electronic databases, 20
Elevator lobbies, 51
Elevators
  avoid using during emergencies, 66
  parking facilities, 52
  recommended lighting levels, 93
Emergency exits, 5, 66
Emergency lighting, 65, 91, 223
Emergency management plan, 64
  incident response, 68–70
Emergency power, 91, 272
Emergency rations, 223
Emergency response plans, 6
Emergency supply kit, 66, 68
Emergency training exercises, 21
Emergency warning system, 65
Employee and visitor numbers, 205–206
Employee entrance, 5, 71, 75
Employee loyalty, 206, 208
Encryption, 196
End user education, 288
Entry control point, 46, *See also* Access control
  data centers, 260
  guards' role, 104, 202
  lighting, 91
Environmental scanning, 296–298
Environmental security design, 30, 79–82, *See also* Crime prevention through environmental design (CPTED)
Equipment replacement, 35, 290–291
Escorting visitors, 102, 103, 104, 202, 205–206, 234
Ethernet over coaxial media, 142
Evacuation planning and procedures, 64–68
Evaluation of security plan, 33, 36–37
Exception handling, 196
Executive protection service, 203
Exercises, 21, 64
Exit signs, 50
Explosive blast protection, 44–45
  standoff distance, 30, 45–46
  window systems, 111–112

F

Facial recognition, 177–178
Facility design, *See* Security construction projects; Security planning and design
Facility factors, determining guard force staffing, 45–46, 204
   hours of operation, 205
   mission of facility, 204–205
   number of employees/visitors, 205–206
   security threat, 205
   size of facility, 205
Facility review, 8
Facility walk-through, 8–9, 12–13
Failure to acquire (FTAR), 191
Failure to enroll (FTER), 191
False acceptance rate (FAR), 193–194
False-alarm rates, 117
False ceilings, 235
False rejection rate (FRR), 192–194
Federal security standards, 82–84, *See also* Sensitive compartmented information facilities
FEMA Emergency Management Guide for Business and Industry, 64
FEMA methodology, security assessment, 15–16
Fences, 87–88
   intrusion detection systems, 120
Filtering, 234
Financial institutions, *See* Banks and financial institutions
Fingerprinting, 178–180, 195, 199
Firearms, 212, 213
Fire drills, 64
Fire protection systems, 23, 117–118, 222
   data centers, 263–265
   design costs, 270
   fire extinguishers, 265–266
   flame detectors, 264–265
   gas suppression systems, 267
   heat detectors, 118, 265
   limited combustible cabling, 263
   smoke detectors, 118, 263–264, 270
   sprinkler systems, 118, 222, 266
Fixed facility checklist, 239–246

Flame detectors, 264
Floods, 22
Fluorescent lights, 92
FM-200, 267
Forensic review video surveillance system, 126
Frame rate, 139
Frame size, 139
Future mapping, 298–300

G

Gaming (casino) video surveillance, 126
   compliance requirement, 123
   use cases, 125
Gates, 88–89
General Services Administration (GSA) security standards, 83–84
Glass
   break sensors, 112–113
   bullet-resistant, 112, 252
   doors, 109
   types of, 112
   windows, 46, 111–112
Global positioning systems (GPS), 63, 299
Government buildings, federal security standards, 82–84
Government security, 227–238, *See also* Sensitive compartmented information facilities
Grooming standards, 211
Guard building, 46
Guards, *See* Security guard force
Gummy fingers, 195, 199

H

h.264 decoding, 134, 141
Handcuffs, 212
Hand geometry, 180–181
HDTV, 126, 138–141, 163
   best use cases, 140
   deployment justification, 143
   HDCCTV, 141–143
   major parameters, 139
   PTZ devices, 131
   standards, 139

video analytics, 140
Hearing impaired persons, 65
Heat detectors, 118, 265
Heating, ventilation, and air conditioning (HVAC) systems, 31–32, 230–231
Heat sensitive cameras (thermal cameras), 135–136
Help desk vulnerabilities, 73
High-definition closed-circuit television (HDCCTV), 141–143
High-definition television, *See* HDTV
High-rise structures, 45, 66
Holdup alarms, 254–255
Holdup buttons, 254
Holdup foot rails, 255
Holistic approach, 6
Homeland Security FEMA Academy, 214
Hosted video solutions, 145–147
Hours of operation, 205
Hurricanes, 21
Hybrid cloud, 148
Hybrid guard force, 209, 219–220

I

Identification badges, *See* Badges
Identification management, 175–176, *See also* Biometrics
In-car and transit video surveillance, 124, 125
Incident response plan, 68–70
Incident response team, 70
Incident threshold, 29–30
Information exchange networks, 20
Information security, *See also* Data centers
    availability, integrity, and confidentiality, 17, 59
    closed storage, 237
    common intrusion tactics and countermeasures, 76–77*t*
    open storage, 236–237
    penetration testing, 72–74
    sensitive compartmented information, 227–238, *See also* Sensitive compartmented information facilities
    shredding, 75

Information technology (IT) related costs, 272
Infrared (IR) flame detectors, 264
Infrared (IR) illumination, 94, 134–135, 138
Infrared sensors
    linear beam, 116
    passive (PIR), 116–117, 119
Infrastructure as a service (IaaS), 146–147
In-service training, 214
INSPASS, 181, 198
Inspections, 202
Instakey, 169
Installation cost, 271
Integration, 152
Integrity, 17, 59, 309
Intelligence Community Directive (ICD) 705, 227
Intelligence sources, 20
Intelligent video search, 150
Intellikey, 168
Interior intrusion detection systems, *See* Intrusion detection systems, interior
Interlaced scanning, 139
Interoperability, video systems, 129, 141, 151–152
Intrusion detection systems, interior, 28, 113–114, *See also* Alarm systems; Closed-circuit television; Video surveillance
    acoustic sensors, 113, 115–116
    balanced magnetic switch, 114
    design costs, 270
    dual-technology fire protection, 117–118
    false-alarm rates, 117
    glass-break sensors, 112–113
    infrared linear beam sensors, 116
    motion-activated cameras, 114–115
    sensitive compartmented information facilities, 235–236, 243–244
Intrusion detection systems, monitoring and responding, 203–204, 219
Intrusion detection systems, perimeter, 118–119, *See also* Video surveillance
    coaxial strain-sensitive cable, 120
    infrared sensors, 119
    microwave sensors, 119–120
    open terrain sensors, 118

time domain reflectometry (TDR) systems, 120
Investigations, 202–203
Ionization type smoke detectors, 264
IP video surveillance, 123, 128–129
    application programming interface (API), 152
    cameras, 131–132
    compression, 129, 133–134
    high-definition, *See* HDTV
    interoperability, 129, 141, 151–152
    lag-free control capabilities, 127
    power over Ethernet (PoE), 128, 131, 133, 136, 163
    progressive scan CCD, 137
    upgrade path, 159
    video encoders, 132–133
Iris recognition, 181–182

J

JPEG, 134

K

Key control, 169
    master locking system, 169–171
Keys, high-tech, 168–169

L

Labor costs, 271
Laminate glass, 46, 112
Landscape security elements, 31, 80–82, *See also* Crime prevention through environmental design
Language issues, 55, 65
Laptop computers, 72, 145
Layers of security, 17–18, 59–62, 114, *See also* Protection in depth
Leadership, 305
    bad leadership, 312–316
    born versus made, 310–311
    dealing with incompetent leaders, 314–316
    defining, 305–306
    effective leadership, 307–308
    going from bad to good, 316–318
    good leadership, 311–312
    importance of security leaders, 308
    leader characteristics, 308–310
    leaders for the future, 318–322
    purpose of, 306
    strategic leadership, 321
Legal considerations, video surveillance, 162–163
Lenses, 137
Lethal Weapons Act 235 Course, 212
License plate capture (LPC) cameras, 134
Life cycle costs, 273
Lighting, 90–94
    color temperature, 138
    emergency, 65, 91, 223
    infrared (IR), 94, 134–135, 138
    luminance, 137
    parking facilities, 52
    recommended levels, 93
    security operations center, 222
    types, 91–92
    video surveillance, 89, 93–94, 137–138
Limited combustible cabling, 263
Loading docks, 53–54
Local crime rate, 205
Locking cylinders, 166
Locks, 110–111, 165, 270
    electric, 110
    electric strikes, 110
    key control, 169
    magnetic, 110–111
    master locking system, 169–171
    types, 166–168
    UL standards, 165–166
Loss-related costs, 274
Loyalty, 206, 208
Luminance, 137

M

Magnetic locks, 110–111
Magnetic stripe cards, 100
Magnetic switches, 114
Maintenance costs, 273
Man bars, 231
Manholes, 30

Man-made threats, 23
Mantraps, 105
Maps, building and site, 65
Maryland State Police, 313
Master locking system, 169–171
Master plan, *See* Security master plan
Matsumoto, Tsutomo, 195, 199
McVeigh, Timothy, 82
Mechanical locks, 166
Mental models, 300
Mentor, 214
Mercury vapor lamps, 92
MESH wireless networks, 140
Microwave sensors, 119–120
Midnight security officer, 13
Mind mapping, 295–296
Mission of facility, 204–205
MITRE Company, 312
MJPEG, 134
Modification loss category, 16
Money clips, 255
Monostatic microwave sensors, 120
Mortise locks, 166
Motion-activated cameras, 114–115
Motion path analysis, 120–121
Movable lighting, 91
MPEG-4, 134
Multibiometric systems, 184–185
Multispectral imaging (MSI), 180, 198
Murrah Building bombing, 82

N

National Industrial Security Programs Operations Manual (NISPOM), 313
Natural barriers, 86
Natural gas detectors, 118
Natural hazards, 4, 21–22
Network defense tests (penetration tests), 70–75
Network operations center (NOC), 260, *See also* Security control center
Networks of security, 20
Network video recorders (NVRs), 144, 250
Night depositories, 250–251
Night vision camera capabilities, 134–136
Nuisance alarms, 105, 117, 223–224

O

Object recognition systems, 148–149, 164
Observation video surveillance system, 126
Oklahoma City federal building bombing, 82
Oleoresin capsicum (OC), 213
Online training, 214
On-the-job training, 214
Open area parking, 52
Open space, 30
Open storage, classified material, 236–237
Open terrain sensors, 118
Operational costs, 271–272
Optical sensors for fingerprinting, 179, 198
Oral reprimand, 217
OSIPS standard, 141
Outsourcing, data center security responsibilities, 262

P

Palm print, 180
Pan-tilt-zoom (PTZ) cameras, 127, 131–132
Parking, 48–50
  access control, 50, 52, 97
  garage facilities, 50–52
  open area, 52
PASS, 266
Passive infrared (PIR) sensors, 116–117, 119
Penetration tests, 70–75
People element of security operations, 6, 25–27, 33, 62–63
Perimeter intrusion detection systems, *See* Intrusion detection systems, perimeter
Perimeter security, 30
  barriers, 86–89
  fences, 87–88
  gates, 88–89
  walls, 89, 228
  CPTED, 79–82, 85–86, *See also* Crime prevention through environmental design
  doors, 109, 235
  federal security standards, 82–84
  fixed facility checklist, 240–241

lighting, 90–94
video applications, 126, *See also* Video surveillance
Permits, 271
Personal identification numbers (PINs), 100, 249, 260
anti-passback programming, 99
Personnel issues, 216–218, *See also* Security guard force
Persons with disabilities, 65, 66
Photoelectric smoke detectors, 264
Physical security, defining, 1–2
Physical security management systems (PSIMs), 152
Physical security planning, *See* Security planning and design
Physical security professional (PSP), 215
Physical security survey, 3
Piggybacking, 75, 105
Planning and design, *See* Security planning and design
Platform as a service (PaaS), 147
Police Scientific Organization, 221
Policies and procedures, 6–7, 30, 34–35, 63
comprehensive building security plan, 84–85
data center layout, 260–261
security organization's strategies, 284
Security Policy and Procedures Manual, 216–217
shipping and receiving, 54
Portable devices, 72, 234
Port Authentication Protocol (802.1x), 130
Positive pressurization, 32
Power over Ethernet (PoE), 128, 131, 133, 136, 163
Power systems
alarm systems, 236
cost factors, 276
emergency backup, 91, 272
security control center, 222
Pressurization, 32
Prevention costs, 274
Privacy issues
biometrics, 190
video surveillance, 162–163
Private cloud, 147

Professional certified investigator (PCI), 215
Professional security certifications, 215
Progressive disciplinary action, 217–218
Progressive scan CCD, 137
Progressive scanning, 139
Project manager, 40
Proprietary security guard force, 206–207
hybrid (proprietary/contract) force, 209, 220
Protection in depth, 17–18, 59–62, 114
access control violation monitoring, 75
basic elements of security, 17, 59
data centers, 260
penetration testing, 70–75
protection plans, 62–64
Protective equipment, 212–213
Protective measures, cost evaluation, 13
Protective service, 203
Proximity card systems, 30, 100, 260, 281, 288
Public arena video surveillance, 124–125
Public cloud services, 147
Public surveillance systems, 129

Q

Quartz lamps, 92

R

Radio equipment, 223
Rate-of-rise heat detector, 118
Ready to exit (REX) technologies, 32
Reception area, as operations center location, 221–222
Receptionist, 81, 86, 101
Recognition-based video surveillance system, 126–127
Recovery planning, 68–69
Recruiting security guards, 216
Refraction type smoke detectors, 264
Replacement costs, 273
Replacement schedules, 290–291
Replay attacks, 195
Reprimands, 217
Request for proposal (RFP), 44
Request-to-exit (REX) systems, 111, 117

Residential video surveillance, 125
Retail video surveillance, 124, 125
Retina scans, 198
Retrofitting, 55–56
Return on investment (ROI), 274–275, 277
Rim locks, 166
Risk analysis, 11, *See also* Security assessment; Vulnerability assessment
    threats and vulnerabilities interrelationships, 16
Risk assessment, bank location, 255–256
Roadway design, 46–47
Rogue employees, 262

S

Safes, 171–173, 248–249
Safety deposit boxes, 247
"Scoop and Go" packs, 252
Screening, 82–83
SD/SDHC memory, 132
Sealing rooms, 67–68
Security assessment, *See also* Threats
    adversary sequence diagram, 36–37
    EASI model, 36
    facility walk-through, 12–13
    FEMA methodology, 15–16
    key questions, 12–13
    security survey, 11–13
    standard questionnaire, 14t
    target identification, 12
    threat identification, 11
    vulnerability assessment, 13–19
Security construction projects, *See also* Sensitive compartmented information facilities
    critical building components, 50, 62
    explosive blast protection, 44–46
    initial access point, 46
    loading docks, 53–54
    parking, 48–52
    penetration testing, 72
    permits, 271
    planning issues, *See* Security planning and design
    retrofitting, 55–56
    review process, 40
    roadway design, 46–47
    security envelope, 44–45
    security team, 43–44
    total system cost determination, 269, *See also* Costs
    windows, 45–46
    working with contractors, 38–40
Security consultants, 280, 281, 284
Security control center, 33
    alarm assessment, 223–224
    communications and monitoring center, 203–204
    contract central station, 113
    cost factors, 276
    design costs, 270
    design requirements, 221–223
    developing an operation, 219–221
    network operations center, 260
    retrofitting, 55–56
    secondary amenities, 223
    staffing, 201, 219–221
Security design concepts, 30–35, *See also* Security planning and design
Security elements
    availability, integrity, and confidentiality, 17, 59
    deter-detect-delay-respond, 19, 61, 247
    people, procedures, and technology, 7, 62–64
Security envelope, 44–45
Security foresight, 293–294
    analogy and metaphor, 300–301
    double–loop learning, 303
    environmental scanning, 296–298
    future mapping, 298–300
    mental models, 300
    mind mapping, 295–296
    wildcard scenarios, 302–303
Security guard force, 201, 219–220
    access control, 99, 103–105
    alarm monitoring and responding roles, 103, 201, 219–220
    armed versus unarmed, 213–214, 287–288
    authority figure, 212
    awareness of security philosophies, 283

balance of manpower and technology, 160–162
bank guards, 255–256
certification, 215
contract, 208–209, 283
contract RFP, 44
costs, 207, 217, 272
entry control, 46, 202
escort responsibilities, 202, 205–206
establishing and size considerations, 201
facility factors and staffing
    hours of operation, 205
    mission of facility, 204–205
    number of employees/visitors, 205–206
    security threat, 205
    size of facility, 205
foot patrols, 104
hybrid (proprietary/contract), 209, 219–220
identification, 211
inspection/investigation duties, 202–203
interviewing midnight officer, 13
layers of security, 18, 63
mission and duties, 202–204
private sector limitations, 104–105
proprietary, 206–207
protective equipment, 212–213
recruitment and personnel issues, 216–218
security planning and design issues, 6, 26–27
training, 208, 214–215
uniforms, 210–211
visitor escort, 104
Security in depth, *See* Protection in depth
Security Industry Association OSIPS standard, 141
Security inspections, 202
Security master plan
    change, 284–288
    consultants and, 280, 281, 284
    contract security relationship, 283
    equipment replacement schedules, 290–291
    security philosophies, 282–283
    security strategies, 283–288
    stakeholder engagement, 280–281
    strategy, 279–280
    technology migration, 281, 288–289
Security networks, 20
Security personnel cost factors, 276
Security philosophies, 282–283
Security planning and design, 25–30, 62, *See also* Security construction projects; Security master plan
    construction aspects, 3–4
    convenience versus security, 4–5, 7
    design concepts, 30–35
    design costs, 270
    evacuation, 64–68
    evaluation, 33
        adversary sequence diagram, 36–37
        EASI model, 36
    holistic approach, 6
    incident threshold, 29–30
    integrating security and function, 7
    making adjustments, 23–24
    people consideration, 6, 25–27, 33, 62–63
    procedures, 27, 34–35, 63, *See also* Policies and procedures
    reviewing the design, 7–9
    SCIF projects, 227
    site planning, 4–7
    site survey, 3
    strategic plan, 27
    technology, 27–29, 35–36
    working with architects, 37
    working with contractors, 38–40
Security Policy and Procedures Manual, 216–217
Security procedures, 27, 34–35, 63, *See also* Policies and procedures
    receptionist knowledge of, 101
Security project team, 3, 43–44
Security retrofit projects, 55–56
Security strategies, security master plan, 283–288
Security survey, 11–13, 201
Security window film, 46
Sensitive compartmented information facilities (SCIFs), 227
    access, 234–235
    alarm requirements, 235–236

closed storage, 237
construction security plan, 227
contractors, 237–238
design
    doors, 228, 242–243
    electrical, 230
    HVAC, 230–231
    perimeter walls, 228
    windows, 228–229
filtering, 234
fixed facility checklist, 239–246
location, 228
open storage, 236–237
shielding, 234
sound masking, 231–233
telephone system, 244–245
TEMPEST, 233
Shelter in place, 66–67
Shielding, 234
Shock wave sensors, 113
Shredding, 75
Signage, 54–55
    data centers, 259–260
    parking facilities, 50–51
Silent alarms, 18, 254–255, *See also* Alarm systems
Single-loop learning, 303
Site planning, 4–7, *See also* Security construction projects; Security planning and design
Smart card systems, 28, 100, 189, 281, 299
Smoke detectors, 118, 263–264, 270
SMPTE 274M, 139
SMPTE 296M, 139
Social engineering, 72–73
Society of Motion Picture and Television Engineers (SMPTE), 139
Sodium vapor lights, 92
Software as a service (SaaS), 145
Sound masking, 231–233
Sound transmission class (STC), 232–233
Spoofing attacks, 195, 197
Sprinkler systems, 118, 222, 266
Staffing central monitoring facilities, 221
Stairwells
    monitoring, 51
    recommended lighting levels, 93

Stakeholder engagement, 280–281
Standby lighting, 91
Standoff distance, 30, 45–46
Strain-sensitive cable, 120
Strategic foresight, 294–295
    analogy and metaphor, 300–301
    double-loop learning, 303
    environmental scanning, 296–298
    future mapping, 298–300
    mind mapping, 295–296
    wildcard scenarios, 302–303
Strategic leadership, 321, *See also* Leadership
Strategic planning, 27, 279–280, 294, 303, 321, *See also* Security master plan
Streaming video, 128, 130
Strong authentication, 175, 197
Suspension, 217
Sutton, Willie, 247
System design costs, 270
System installation costs, 271

T

Tailgating, 75, 105, 196
Tamper-resistant solutions, 196
Target hardening, 79
Target identification, 12, 33
Teams, 3
    good leadership, 311–312
    implementing change, 285–286
    incident response, 70
    security project, 3, 43–44
Technology, *See also specific technologies*
    ASIS resources, 35
    deterrence capability, 28–29
    future mapping, 298–300
    IT-related cost, 272
    layers of security, 63
    life-cycle replacement, 35
    manpower and, 160–162
    migration, 281, 288–289
    planning issues, 27–29, 35–36
    retrofitting, 55–56
    security program integration, 33
    upgrade path, 159
    working with contractors, 38–40
Teller cash recyclers, 251

TEMPEST, 233–234
Template database attacks, 195
Template storage, biometrics, 188–189
Termination, 217
Terrorism, 82
Testing network defenses (penetration tests), 70–75
Testing response plans, 70
Theories of action, 300
Theory in use, 300
Thermal cameras, 135–136
Third-party security assessments, 262
Threats, 8
    CPTED approach, 85–86
    factors determining guard staffing, 205
    man-made, 23
    natural hazards, 21–22
    threat definition, 19–20, 33
    threat identification, 11, 27
Three-dimensional contactless finger imaging, 180, 198
Three-dimensional ultrasonic imaging, 180, 198
Three-factor security, 176
Time-domain reflectometry (TDR) systems, 120
Token-based authentication, 175–176
Tool-resistant safe class TL-15, 171, 248–249
Tornadoes, 21–22, 67
Total cost of ownership (TCO), 275, 277
Total system cost, 269, *See also* Costs
Traffic monitoring, 46
Training
    emergency exercises, 21
    evacuation activities, 64
    lethal weapons, 212
    security guards, 208, 214–215
Transportation video surveillance, 124, 125
Trash, 74
Truck bomb threat, 85

U

Ultrasonic imaging, 180, 198
Ultrasound sensors, 179, 198
Ultraviolet (UV) flame detectors, 264–265
Underground parking, 48, 50

Underwriters Laboratories (UL) standards
    central station staffing (UL 1981), 220–221
    locks (UL 437), 165–166
    safes, 171–173, 248–249
Uniforms, 210–211
Union representatives, 280
Upgrading video systems, 159
Urban video surveillance, 124
US Marshals Service Building Security Study, 82–83

V

Varifocal lenses, 137
Vaults, 173, 236, 247–248
Vehicle authorization, 3
Vehicle barriers, 31
Vehicle parking, *See* Parking
Vehicle traffic monitoring, 46
Vehicular circulation patterns, 31
    roadway design, 46–47
    signage, 54
Vein pattern recognition, 182–183
Verification transaction, 187
Video analytics, 125, 140, 148–149
Video compression, 129, 133–134
Video content analysis (VCA), 120–121, 125, 149
    intelligent search, 150
    video synopsis, 125, 150–151
Video encoders, 132–133
Video management systems (VMSs), 144, 164
    cloud computing, 144–148
Video recording systems, 143–144
Video streaming, 128, 130
Video surveillance, 123–124, *See also* Cameras; Closed-circuit television; IP video surveillance
    alarm assessment, 223–224
    analog systems overview, 127
    ATM transactions, 250
    bank systems, 253–254
    central station design, 221–223
    central station staffing, 219–221
    cloud computing and hosted solutions, 144–148

compliance requirement, 123–124
digital video recorders, 143–144, 222, 250
future of, 163–164
high-definition, *See* HDTV
illumination, 89, 93–94, 137–138
integration, 152
interoperability, 129, 141, 151–152
intrusion detection systems, 120–121
layers of security, 18
manpower and, 160–162
motion-activated cameras, 114–115
parking facilities, 51
privacy and legal considerations, 162–163
smart video, 121
system classifications, 126–127
system selection and deployment, 152–158
upgrade path, 159
use cases, 124–126
Video synopsis, 125, 150–151
Visitor and employee numbers, 205–206
Visitor entrance, 81, 86
Visitor management, 101–103
    escort, 102, 103, 205–206, 234
    receptionist role, 101
    software, 64, 102
    temporary badges, 101–103
Visitors' log, 101, 103, 234–235

Visual tools and techniques, 294
Vulnerability assessment, 11–16, *See also* Security assessment
    asset value rating, 15
    developing, 16–19
    penetration testing, 72
    target identification, 12, 33
    US Marshals Service study, 82–83

W

Walls, 89, 228
Warehouse style structure, 45
Warning system, 65
Water flow sensors, 118
Water sprinklers, 118, 222, 266
Weakest link, 18, 104
Weapons, 212, 213
Wildcard scenarios, 302–303
Windows, 45–46
    alarm systems, 112, 235
    CPTED, 82
    glass-break sensors, 112–113
    sensitive compartmented information facilities, 228–229
    types of glass, 112
Wired glass, 112
World Trade Center attacks, 82
Written reprimand, 217